MORE GOOD NEWS

DAVID SUZUKI
HOLLY DRESSEL

MORE GOOD NEWS

REAL SOLUTIONS TO THE GLOBAL ECO-CRISIS

 David Suzuki Foundation

 GREYSTONE BOOKS

D&M PUBLISHERS INC.

Vancouver/Toronto/Berkeley

Greystone Books
An imprint of D&M Publishers Inc.
2323 Quebec Street, Suite 201
Vancouver BC Canada V5T 4S7
www.greystonebooks.com

David Suzuki Foundation
219–2211 West 4th Avenue
Vancouver BC Canada V6K 4S2

Cataloguing data available from Library and Archives Canada
ISBN 978-1-55365-475-9 (pbk.)   ISBN 978-1-55365-625-8 (ebook)

Cover design by Ingrid Paulson
Text design by Heather Pringle
Cover illustration © Masterfile
Printed and bound in Canada by Friesens
Text printed on 100% post-consumer,
acid-free, FSC-certified paper
Distributed in the U.S. by Publishers Group West

We gratefully acknowledge the financial support
of the Canada Council for the Arts, the British Columbia Arts Council,
the Province of British Columbia through the Book Publishing Tax Credit,
and the Government of Canada through the Canada Book Fund
for our publishing activities.

Mixed Sources
Cert no. SW-COC-001271
© 1996 FSC
FSC

CONTENTS

ACKNOWLEDGMENTS

As in any work of this kind, it would be impossible to thank the scores of people who gave selflessly of their precious time and expertise to answer calls and queries, to send information, to find books. At the very beginning of this project, Kristen Werring, a wonderful volunteer at the David Suzuki Foundation, lent her talents to making sure our websites and references were all existent and up-to-date. We want to thank our publishers for their patience in dealing with a large and unwieldy early manuscript. Thank you so much Rob, Nancy, and especially our editors, Barbara Tomlin and Iva Cheung. Barbara's kindness, availability, and deep understanding of the book's intent, as well as her clear interest in the issues involved, made the difficult work of refining and editing a joy. Iva's brilliance at ferreting out embarrassing mistakes or omissions, as well as clarifying thoughts, was just as deeply appreciated. We are very grateful for being the recipients of the gentle professionalism of these editors. David wishes to thank Tara Cullis, "for all her support in giving me time to write and for her input on the forestry chapter. My right-hand person, Elois Yaxley, keeps the forces of

1

entropy at bay in our office." Holly Dressel would like to thank Ryan Young, who threw references her way, loaned books with abandon, and was also a consultant on many key issues. James Latteier read Holly's earliest manuscripts and helped with corrections and especially morale. She would also like to mention the unwavering support she has been given over this period by Peter Brown, Elizabeth May, and Christine von Weizsaecker, as well as by her friends Stuart Myiow Jr. and Sr. Holly's daughter, Thea Toole, and daughter-in-law, Amy Stolecki are local environmental activists and provided invaluable primary research on sludge and other solid waste issues. And most of all, Holly "is grateful for my two grandsons, Parker and Owen, who provided desperately needed cookie, cartoon, and hide-and-seek breaks." It is to them and to all the other children, of today and tomorrow, that we both offer this work.

INTRODUCTION

Are there real alternatives to the human activities that undermine the biosphere's life-supporting systems—alternatives that are not just about recycling and riding the bus but that represent substantive shifts in our perspectives, values, and goals?

That's what we set out to discover when we began work on *Good News for a Change* back in 2000. At that time we worked hard to discover examples of the changes the world needed and to outline real sustainability criteria: principles that people could follow to devise and recognize solutions to environmental problems. Since then reports of ecological collapse have increased in number, capturing public attention and urging people and governments to more quickly adopt the existing solutions. We have now combined those original sustainability principles with new examples in order to impel faster movement along a clear path.

Although people must continue acting to save the planet's failing ecosystems, it is just as vital to distinguish actions that are theoretically helpful from actions that truly benefit natural systems. Sometimes exciting "green solutions," such as biofuel production, can cause more harm than good. Conservationists and concerned citizens everywhere must use clear criteria to judge what actually keeps a particular ecosystem in balance, given each one's staggering complexity. Unlike most books about environmental solutions written at that time, ours was *not* enthusiastic about agricultural biofuels, because growing fuel on finite agricultural land to feed infinite human desires for energy did not (and still does not) fit our sustainability criteria. Within only two crop seasons, subsidies favoring biofuels began to destabilize food prices and destroy ecosystems around the world.

Our recommendations now, as in the first edition of this book, are based on science, which acknowledges that natural systems are not part of human culture; we are part of natural systems. The ecosystems that keep us alive pay no attention to human aspirations, needs, or laws. A nation might work out the most elegant carbon-trading program imaginable, one that takes into account all the necessary political and economic realities. Yet if that program doesn't lower greenhouse gas emissions fast enough, the climate won't pay any attention to such efforts. That hard fact means that we must try to understand what forests and water bodies really need in order to maintain their ability to regulate this planet's atmosphere and climate.

As well as using the best available scientific research, we have turned to the most established experts on the biosphere's needs— the local people who depend upon, live within, and therefore have paid very close attention to natural ecosystems. It is in the work done by scientists and local people that the most hope for saving the planet is growing. More scientists are recognizing natural complexity and, with it, the fact that humans do not understand ecosystems well enough to manage them sustainably. So far science hasn't been able to isolate or understand most of the components of healthy ecosystems, let alone comprehend how they are interconnected. The entire science of conservation is learning humility, and that includes turning to history to see what human groups have done in the past that was demonstrably successful. Today academic and governmental experts are acknowledging that the local, traditional, and aboriginal peoples who still live within the planet's remaining functioning ecosystems are most able to tell us what these systems require. These groups talk about "respect," "restraint," and "listening to the land." Research is confirming that this is very good advice for saving the biosphere—that people have a better chance of learning to manage human behavior than they have of managing the infinitely linked ecosystems that support us. This is a lesson that both authors of *More Good News* have learned firsthand.

When Holly was a child, her grandfather kept several hives of bees on his small farm in rural Ohio. Like any little kid, she was afraid of being stung and during summer visits would avoid the large numbers of bees buzzing in the backyard under the long grape arbor or clustering at the lip of a big birdbath. Holly's grandfather didn't approve of her fear, so one summer afternoon, when she was five or six, he taught her how to gently approach the bees getting drinks from the birdbath, moving slowly and either talking or thinking calm, friendly thoughts. In ten minutes she was putting her hand near the water and letting the honeybees climb on, a dozen at a time. They tickled, but they didn't sting. Grandpa Dressel had great contempt for beekeepers who were afraid of their hives. "If you move gently, if you let them know you mean them no harm, they'll never sting you," he said, and he demonstrated this to Holly years later when moving two huge swarms. This lesson about mutual respect and communication between humans and nature, to the benefit of both, is one that any traditional hunter, farmer, or fisher can confirm.

When David was a young man, he was galvanized by Rachel Carson's 1962 book, *Silent Spring*, and swept into the exciting new environmental movement. David thought humans were taking too much from their surroundings and putting too much waste and toxic material back. He believed we could solve these problems by establishing departments of the environment; passing laws to protect air, water, soil, and species; and applying some promising technologies. Over time, however, David saw that even respected scientific experts didn't know enough about natural systems to manage human impacts; another approach was needed. In the late 1970s he had an experience that suggested what this approach might be.

He was documenting the battle over forests in Haida Gwaii, where logging provided jobs for many Haida. So he asked Guujaaw, a Haida leader, why he opposed logging when it brought economic benefits to his community. Guujaaw answered, "Because when the trees are gone, we'll just be like everybody else." To David this simple statement opened a window on a radically different way of seeing the world. Haida do not see themselves ending at their skin or fingertips. Being Haida means being connected to the rest of creation: the water, air, fish, and trees make the Haida who they are. David says, "They taught me that Earth is our mother and that we are created out of the four sacred elements: earth, air, fire, and water. They are grateful to their 'relatives'—other species that provide all we need to survive."

Understanding our connection to other living things is key when working toward a different world. It is all of life, the biosphere, that allows us to live and flourish, that makes any human economy possible. Protecting rich and intact ecosystems should be our highest priority. In our research for this book we found, as we did when we wrote *Good News for a Change*, that many people recognize the need for functioning ecosystems. Around the world, local and traditional methods of conserving habitat and watersheds are gaining ground, and at the same time political and economic strategies are reducing greenhouse gas emissions, improving agricultural methods, and helping to control population and promote green lifestyles.

Because political will is so important, the first chapter in our updated story of good news, "Viva la Revolución," is about how people are using political means to get a handle on threats to the planet. This chapter outlines the sustainability criteria contained in the first edition of our book and expands on the importance of approaches that mimic nature, are flexible, allow for the possibility that an underlying assumption about a natural system might be wrong, involve bottom-up democratic management, and set lofty and long-term goals. As well as revisiting the ideas of William McDonough and reviewing the success of The Natural Step movement, we talk about Scheer's Law in Germany, which is revolutionizing the world's political approach to climate change, and the new Bolivian constitution, which may do the same for natural systems.

Chapter 2, "Using Coyotes to Grow Grass," is about the importance of maintaining biodiversity. We again discuss the Adirondack Park model and Allan Savory's Holistic Management approach, because these remain the foundation for the growth of 4c (Cores, Corridors, Carnivores, and Communities) movements around the world. This chapter also contains some of the most heartening news about the spread of effective wildlife and biodiversity preservation.

Chapter 3, "Avoiding Venus," is almost all new and focuses on how to effectively address climate change. We carefully analyze the many suggestions, most of them economic, political, and technical, for reducing greenhouse gas emissions. Some, like public transport and energy conservation, are very promising, whereas others, like first-generation biofuels and geoengineering, could turn out to be worse than doing nothing. We also assess a variety of renewable energy options and offer a way to judge what will really clean up the air and stabilize the atmosphere.

Chapters 4, 5, and 6 focus on the forest and water ecosystems so central to life on Earth. In Chapter 4, "Listen for the Jaguar," we show how recent ideas for local forest management have spread across the planet. In Chapter 5, "A River Runs Through It," we recount the serious challenges affecting the sublimely necessary resource of fresh water and outline the many ways in which present water management must change. We also offer uplifting examples of how very quickly watercourses can heal when we let them flow. In Chapter 6, "The Mother of All," we describe the grave plight of our acidifying oceans, which are absorbing gases from the air and gathering in all the wastes of human activity. Although awareness of this crisis is growing and some real heroes are at work, we must do more to carry the rule of conservation law into international waters.

Thankfully we have a happier story to tell in Chapter 7, "Baking a Sustainable Pudding," where we describe the new organic and locavore food initiatives that have spread all over the world. The power of industrial agriculture remains a challenge but one that is being met as national and UN agencies turn away from agribusiness and toward "multifunctionality"—small, diverse landholdings managed by local people to feed themselves and others.

Our last chapter turns to where most of us live. In "Green Jobs in the City" we look at the new movement to "green" every kind of job and create more livable, sustainable cities. We also tackle two of humanity's biggest concerns: our growing population and our obsession with financial wealth. How many people can the planet support, and can we keep acting on the assumption that money will buy all these people happiness? The answers may not be welcome to everyone, but they offer humanity simple and achievable solutions. They depend on our species' proven, but not always practiced, ability to cooperate and to share.

In this book we have tried to provide a blueprint for healing a severely damaged planet. We have attempted to look problems squarely in the eye and to assess solutions with skepticism and suspicion as well as with enthusiasm and hope. We can't deny that in the last several years the planet's ecological crises have grown. That makes it all the more urgent that we understand the examples provided here and follow the approaches outlined for healing the biosphere and living within the limits of nature's bounty.

We hope reading *More Good News* will provide readers with inspiration and energy for this hard, exciting, and supremely fulfilling work.

Note: As in our former books, we often use the terms "First World" and "Third World" to describe developed and developing nations. Our contacts in developing countries have expressed a preference for these terms, as they imply another approach to apprehending and managing the Earth—a different kind of World.

1

VIVA LA REVOLUCIÓN

POLITICAL WILL

TRIUMPH

What is the intention of [this] movement? If you examine its
values, missions, goals, and principles, and I urge you to do so,
you will see that at the core of all [these] organizations are two
principles, albeit unstated: first is the Golden Rule; second is the
sacredness of all life, whether it be a creature, child, or culture.

Paul Hawken, Blessed Unrest[1]

Concerns about environmental problems have deepened since the
first edition of this book was written, and for good reason. In the
past few years the acidification of our oceans, the loss of glaciers
and ice caps, the destruction of forests and agricultural land, and
the increased instability of the planet's weather have only gotten
worse. Fortunately human responses to this state of crisis have also
mushroomed. Every day more people are working to defend and
restore ecosystems, share resources more equitably with local people, and assess what constitutes real human happiness and purpose,
and some have achieved some heady successes on the level of local
ecosystem health.

Back in the 1960s, '70s, and '80s, many groups and individuals were concerned about toxic waste dumps, increasingly despoiled forests and farmlands, dead zones in the oceans, and the discouraging diebacks of once-common wildlife like striped bass, wild turkeys, bald eagles, and sea otters. Today all four of those endangered species are recovering and even thriving, thanks to remarkable efforts from individuals, combined with tough protective legislation. All over the world laws have prevented toxic waste dumps from growing, and now an international treaty is in place to eliminate "the dirty dozen," the twelve most dangerous chemical compounds in existence.

Almost every country in the world now has an environmental department, and many municipalities have appointed environmental or sustainability commissioners. People like David Day in Calgary and Sadhu Johnston, Chicago's chief environmental officer, can exercise authority within all branches of city government, "making sure the green tint runs to the core."[2] There is barely a large supermarket left in the industrialized world without an organic food section, and there are thousands of fully organic markets right across our continent.

In many parts of the world it's becoming common for farmers to receive grants to reserve land as wildlife or wetland habitat; poor countries like Bolivia and Ecuador are taking control of their water and forests; and although the battle for the Amazon still rages, some truly progressive legislation has slowed the decimation of remaining tracts of boreal forest in Canada. This legislation does not try to replace wild ecosystems with monocultured plantations but is turning the forests over to the indigenous groups that know best how to manage them. The first clear understanding of sustainable habitat protection is the "Cores, Corridors, Carnivores, and

Communities" movement that was just gaining steam in certain parts of North America a few years ago. The 4cs have now become the accepted standard of wilderness and habitat remediation. Meanwhile the Slow Food and locavore movements are delighting gourmets around the world by ensuring that delicious foods are produced in ways that protect soil, air, water, and wildlife.

The list of accomplishments goes on. The growth in every kind of sustainable building—from solar- and wind-powered high-rises to tiny, 84-square-foot "eco-houses"[3]—has been nothing short of phenomenal. Hundreds of books and thousands of websites are analyzing the pros and cons of a bewildering new array of energy sources. Expanded geothermal and wind energy technologies are becoming more sophisticated and available every day, and new tidal and wave generators are generating relatively clean power in New York's East River and off the coasts of Portugal and Washington State. As for solar, the safest and most basic energy source of all, large-scale technologies that go beyond panels that heat water or generate small amounts of electricity are now in place, using heliostats, solar troughs, parabolic dishes, and solar power towers. Self-energizing cars that feed energy back into the grid are finally coming onto the market.

The last big push to save the environment through such technologies was slowed by an economic downturn in the late 1980s that seemed severe at the time but was far less so than the one that started in 2008. After the heady moment of the Earth Summit in 1992, the economy suddenly became far more important than saving natural systems. Ecological concerns were pushed aside for more than a decade, spawning runaway, unregulated financial growth and industrial development. Today, with the unmistakable signs of systemic ecological breakdown in evidence—the loss of glaciers, the collapse of fisheries, and the alarming changes in the chemical makeup of the planet's atmosphere and oceans—it is impossible for intelligent people to think that the environment and the economy are unrelated or that we can tackle one without taking care of the other. People have recognized that there can be no economic wealth without a productive biosphere. They are taking action in response to this fundamental truth.

YOU SAY YOU WANT A REVOLUTION?

What if our economy were organized not around the lifeless abstractions of neoclassical economics and accountancy but around the biological realities of nature?

Paul Hawken[4]

Inspiration for some of the most fundamental mind-shifts in recent years can be found in the work of a few seminal people, including architect and designer Bill McDonough, the former dean of architecture at the University of Virginia. He says that in the West we've now had two industrial revolutions. The first was about resource extraction and money. The second is happening now, and it's about resource conservation and values. McDonough says that if you were to articulate the First Industrial Revolution as a design assignment to a class of students, you'd have to say something like this: "Could you design a system that pollutes the soil, air, and water; that measures productivity by how few people are working; that measures prosperity by how much natural capital you can dig up, bury, burn, or otherwise destroy; that measures progress by the number of smokestacks you have; that requires thousands of complex regulations to keep people from killing each other too quickly; that destroys biodiversity and cultural diversity; and that produces

things that are so highly toxic they require thousands of generations maintaining constant vigil, while living in terror?"[5]

This is undoubtedly not the system that the people working
toward the First Industrial Revolution at the end of the eighteenth
century meant to create, but it's a pretty good description of the
one we got. McDonough says an industrial system is now being
designed on the basis of a new paradigm, and he can describe how
it will work: "The Next Industrial Revolution introduces no haz-
ardous materials into the air, water, or soil; measures prosperity
by how much natural capital we can accrue in productive ways;
measures productivity by how many people are gainfully and
meaningfully employed; measures progress by how many buildings
have no smokestacks or dangerous effluents; does not require regulations whose purpose is to prevent us from killing ourselves too
quickly; it produces nothing that will require future generations to
maintain vigilance; and celebrates biological and cultural diversity
and solar income."[6] That's the revolution that's starting up around
us and that marks a real, systemic change in every level of society,
all over the world.

In the current, clearly failing economic system, we have typically been required to make hard, "either-or" choices: a new factory
and the jobs it will provide, or a clean river? Grazing land for ranchers or local tribes, or habitat for wolves and tigers? Pristine rivers
and forests in northern Alberta, or a devastated, toxic moonscape
that brings immediate cash? Activists get bitter when the crudest,
most short-term economic human needs win out over the long-term
necessities of the biosphere almost every time. They see that the net
result all over the world of separating economic from environmental or social concerns has been dirtier rivers, lousier neighborhoods,
dwindling numbers of wild animals, and a frighteningly altered

atmosphere. These purely economic considerations—the bias to create factory or infrastructure jobs and to manage nature for immediate, short-term benefit—have preoccupied us ever since the First Industrial Revolution, when people, especially in the Western world, first began to slice nature into its component parts and to see it as a machine to be managed piecemeal for human use.

Throughout the long human past it has been frightening to depend entirely on nature's erratic bounty, but we've done more deeply systemic damage to natural ecosystems in the past three hundred years than in the preceding hundred thousand. So, given that we have managed to subdue nature quite effectively and increase our own numbers and dominion beyond belief, the question people have been asking themselves is whether we could have some of *both*: paying, secure jobs and industries *and* a decent base of natural systems to support them, instead of all one or all the other. And this new attempt to combine our technical gifts with our expanding knowledge of how the planet really works is what the Second Industrial Revolution is about.

Paul Hawken, former mainstream entrepreneur, founder of the gardening company Smith & Hawken, and author of *Natural Capitalism, Blessed Unrest,* and *The Ecology of Commerce,* says that what's most exciting about the great variety of groups involved in this Second Industrial Revolution is the fact that "they agree, to an unprecedented extent, on an extremely similar vision of the future." Hawken points out that one of the most remarkable developments of the late twentieth century is the "tens of thousands of NGOs [non-governmental organizations]" that have spontaneously arisen in just the past few years. These organizations address everything from social justice and population to corporate and electoral reform, environmental sustainability and renewable energy.[7]

The NGOs Hawken talks about keep in touch especially on the Internet, by phone, and at meetings all around the world. They work at local, state, and international levels, and they've all been moving toward a vision of a truly sustainable world. They're writing down their ideas, says Hawken, "creating conventions, declarations, lists of principles, and frameworks that are remarkably in accord," and then they pressure business and government to adopt them. But most important, according to Hawken, is his realization that, "never before in history have independent groups from all around the world derived frameworks of knowledge that are [so] utterly consonant and in agreement. It is not that they are the same; it is that they do not conflict. This hasn't happened in politics, not in religion, not in psychology; not ever."[8]

Besides writing books that have changed the social and environmental focus of many large companies, Hawken brought The Natural Step (TNS) to North America. This ingenious system was created by a Swedish doctor, Karl-Henrik Robèrt, who realized that industrial toxins were killing his young patients. It sets up four simple conditions that must be met if a society is to achieve environmental health and sustainability. The first requires that as little material as possible from underneath the Earth's crust be brought to the surface (since so many of these materials are poisonous to life). The second requires that humans not subject natural systems to overly large concentrations of any of the materials that they produce. The third stipulates that the Earth's natural systems must not be systematically depleted through overharvesting, displacement, or other manipulation.

The fourth condition, Hawken says, is the one that causes his mostly corporate audiences to balk and underlines the difference between their attitudes and those of the activists working for a

New Industrial Revolution: "Without social justice and fair and equitable distribution of resources, there can be no such thing as sustainability." Hawken says. "One of the most humorous aspects of teaching The Natural Step in corporations is that when you come to the Fourth System Condition ... businesspeople go ballistic. They think it is socialist, communist, the nose of the leftist camel slipping under the tent. Literally, some are repulsed by it. We are in a country that was founded on 'liberty and justice for all,' and if you raise that issue in the business community, some executives will fall off their chairs."[9]

Social justice is a foundation of any sustainable management system simply because if one group involved in, say, growing timber, gets too many benefits and another gets too many disadvantages, those profiting will overexploit and those being left out will revolt. Both are common symptoms of unsustainable practices, as the resource being managed becomes a bone of contention instead of a steady source of renewable support. What is interesting is that the incredulous businesspeople, when they play the role of seriously designing workable, closed societies, nearly always end up demanding the fourth TNS condition of bottom-up equality they had initially rejected. Hawken concludes, "In small groups, with appropriate goals and challenges, we all know the right things to do. As a society within the world of corporate capitalism, we are not very bright."

Over the hundreds of thousands of years we've been on this planet, human beings have developed many values and many ways to live. The current paradigm of bringing every region and every political group on the planet under the aegis of a neoliberal economy directed by private corporations, an idea that has grown quite naturally out of the precepts of the First Industrial Revolution, has spread like a proselytizing religion throughout the world.

It has assumed that its concept of values (money) and lifestyle (ever-increasing security and luxury) is the best, a kind of pinnacle of social evolution, and that people will all be better off and happier if they adopt these values, even if some of those people reap a lot more of the benefits than others. But like any dogma, it has fatal flaws, the most glaring being that all its assumptions rest upon ignoring this planet's physical laws. The finite amount of air, water, and soil on Earth means only so much luxury and security can be supplied to the ever-growing numbers of humans before these material comforts run out.

What the New Industrial Revolution of the twenty-first century is already accomplishing, besides saving species and legislating against pollution and resource waste, is building a new system of values based not on ridiculous Scrooge McDuck fantasies of unlimited riches but on the simultaneously limited and miraculous realities of physics. And it is these realities that are teaching us all about the real joys of being alive.

DOUBLE INDEMNITIES, DOUBLE DIVIDENDS

Companies that pursue these natural principles are not only more profitable; they're more fun. All of a sudden they have intellectual challenges again, [and they're] rewarding employees who eliminate waste, empowering their workers to be responsible.
Hunter Lovins, Rocky Mountain Institute[10]

We're used to the fact that supporting First Industrial Revolution development requires highways, harbor facilities, factories, and other expensive infrastructure that can only be built with heavy outside investment. We also assume that such projects will require expensive technologies and machinery and that they'll need continuous public funding in the form of subsidies, tax breaks, special grants, and rebates. But in the New Industrial Revolution we're learning that when something works in harmony with our planet's natural cycles and systems, it can bypass many of the rules that conventional economic experts claim are essential to industrial function. Of course really sustainable development could use some economic incentives at the outset when it's competing with entrenched mass products that do have subsidies. But in general truly sustainable systems don't need the heavy capital investments demanded by, say, biotech crops or oil refining and definitely shouldn't need constant artificial support once they get going, the kind of heavy tax support the automotive and industrial farming industries continue to require. In fact if an established industry or product does demand such support, that's the first clue that it might not be truly sustainable.

Second Industrial Revolution development also doesn't have hidden liabilities that dump pollution and energy costs on society at large, greatly increasing the original capital expenditure that has to be made by people who usually don't even share in the benefits. When technologies or management practices work in concert with physical reality, they always provide a great deal more than one benefit. Coming to a decision to save watercourses, for example, by raising hogs in a sustainable, organic manner, not only prevents dangerous antibiotic-resistant bacteria and new strains of viruses from contaminating surface water but returns pastureland to a greater state of diversity, enabling it to support more animals that can interact out in the sunshine, while providing a stable, long-term income for more people than the current mechanized, industrial feeding operations ever do.

First Industrial Revolution principles are so at odds with natural requirements that they can cause illness and death. For instance, by operating one central packaging facility—the ideal of business efficiency—the mammoth Maple Leaf Foods plant in Toronto became the source of the deadly 2008 listeriosis outbreak. Industrialized food production and processing means that one dirty machine, thousands of miles away from millions of consumers, can spread death across a huge area.

Industrialization also co-opts government, because wealthy industrialists lobby aggressively for, and usually get, relaxed standards and testing. Huge industrial facilities, whether raising or processing food, respond to economic needs, not natural system needs, and inevitably become too large to control. As one reporter put it, following the listeriosis deaths, "The gutting of the inspection system is part of the ongoing deregulation of industry and the program of budget cutting that has been pursued for the past two decades by all governments, federal and provincial... The fact that the new food inspection system was designed by the Liberal government and then enacted by the Harper Conservatives should come as no surprise. Both parties are instruments of big business who pursue policies that put the interests of private profit before human need."[11]

The way we raise, process, and distribute food industrially and emphasize quantity and economic values over quality and health or ecosystem values is a perfect example of an unsustainable management system. Currently most of our food is raised under the old industrial paradigm, which has hidden pollution from fossil fuel–based chemical inputs, excessive water use and soil contamination, and long-distance transport. This makes industrial farming one of the biggest contributors to climate change. Even if owned or operated by families, the typical industrial hog farm, for example, uses a skeleton crew of very few employees in charge of producing fifty thousand pigs a year; the pigs must be fed hormones and antibiotics, or the stresses caused by overcrowding will keep them from eating properly. These operations are vertically integrated, so the farm will be closely connected to a slaughterhouse and exporting facility owned by a large multinational corporation. Because of government subsidies on everything from production insurance to transportation costs, that corporation might show a profit, which is interpreted as a rise in a country's GDP. But other very serious costs are never taken into account. One of them is increased and emerging diseases in the animal and human populations.

Epidemiologists have been waiting for a so-called swine flu pandemic ever since we began raising hogs this way, creating perfect conditions for new viruses and antibiotic-resistant bacteria to develop. The recent increase in community-acquired Methicillin-resistant *Staphylococcus aureus* (MRSA), which can lead to an often fatal form of flesh-eating disease, is now proven, like H1N1, to arise in the industrial farm setting.[12] Industrial farming practices lead to further contamination of rural soil and watercourses by the high concentrations of manure and undesirable chemicals in the manure, such as hormone mimickers from the antibiotics and poisonous pesticide residues from the feed. Both the Canada and American medical associations have for years demanded a moratorium on industrial hog farms, but North American governments favor this powerful industry, which has spread to Mexico and the tropics, where it will become even more dangerous. How many more warnings do we need to see that the industrial method of raising food is unsustainable?

If the tax funds that currently prop up outdated practices with free research, installation, and transport breaks were used differently, as they increasingly are across Europe, agriculture would be not only safer and far cheaper to maintain over the long term but actually more productive and much more economically viable for local farmers. Farms would have to get smaller again and have more people taking care of fewer animals, a proven method of increasing physical health and economic benefits for rural farmers and even urban consumers. And if we're serious about letting the market decide, we must *equalize* our system to allow small producers to get the same kinds of benefits, access, and support as the huge multinational corporations. Better yet, as in Germany, North American agricultural policies should gradually cut any government subsidies to any farms that dump petrochemical wastes and depleted soils and water tables on future generations.[13] Our governments should feed that money into organic operations, helping farmers to switch to sustainable practices, and then gradually phase out subsidies and let the market work properly.

The easiest way to identify Second Industrial Revolution practices is to look for double dividends and an absence of indemnities, especially over time, and organic farming has been a prime example of how that works. With almost no help from subsidies and in spite of many onerous regulations expressly intended to destroy it, organic farming has still grown, continent-wide, at the remarkable rate of 20 percent a year for more than two decades, supporting more and more families in the process, providing millions with purer food, and conserving land and water. Imagine what it could do if we simply removed the legislation favoring its unsustainable counterparts!

Second Industrial Revolution proponent Bill McDonough says we need to ground ourselves in the measurable reality of how natural systems work. The planet Earth, he explains, "represents chemistry; and the sun, physics. When the two get together, an energy source is created—water and soil under the solar flux, creating a kind of photosynthetic energy cell."[14] This physical-chemical energy generates plants and animals on the thin surface of the planet—what is termed "solar income"—so generously that it produces more than the overall system needs. "Solar interest" is the excess fruit that feeds massive numbers of insects, the birds that eat the excess insects, or the millions of fish fry that nourish other fish. That excess can be harvested by humans and other life-forms for food energy, without harming the ability of the solar-powered system to keep on producing. But if any species, say a particular fish or tree, for example, is consumed faster than it can reproduce, then we aren't living on solar interest but on solar capital. We're depleting the capital we should be keeping in the "bank," which will produce the income needed to support the Earth's children in the future.

We need, McDonough argues, to pattern our industries on nature, what is termed "biomimicry." That means preventing waste by limiting any sort of manufacturing or consumption that causes it. If any waste is created, it has to be nonpolluting so that it can serve as "food" for natural or industrial systems, just like dead leaves or muddy water in nature. If your waste can't be safely absorbed by natural systems or fed back into industry easily, that's the proof that your methods don't accord with nature's, that you're losing part of your investment and causing damage. Indigestible wastes like toxic chemicals or heavy metals also destroy natural systems' ability to produce more products in the future.[15]

The Rocky Mountain Institute in Colorado is the ecological think tank that has spent the last thirty years grinding out the nuts and bolts of the New Industrial Revolution—the curly light bulbs, solar panels, and hybrid cars that are in mainstream markets today. The institute's founders, the lawyer–inventor team of Hunter and Amory Lovins, have been able to illustrate very clearly what they call the "diseconomies" of large-scale energy, food, and industrial production. By mimicking nature, which works in multiple ways on small and extremely varied scales, people like McDonough and the Lovinses have been able to evolve whole new technologies: finding ways to make cars and machinery that use nontoxic building materials like silica and carbon, and figuring out technologies that depend on renewable solar, wind, or hydrogen power. These have been in development for decades, as we all know, but they're finally being allowed to hit the market.

Even though cars, air conditioners, computers, and so on seem expensive, their retail prices don't begin to reflect their true costs, even to those of us who can afford them, because the First Industrial Revolution "externalized" these costs, dumping the lion's share on society at large. And in fact those who suffer the most from this externalization, poor people exposed to toxic chemicals, for example, are generally those who have the least opportunity to benefit, an injustice that proponents of the First Industrial Revolution defend themselves against by denying any intrinsic connection between their activities and notions of morality.

The senior partner in the First Industrial Revolution process, science itself, is becoming increasingly at odds with its earliest incarnation—that is, the reductionist form it first took that mechanically separated ethics, physics, mathematics, politics, history, and so forth. For several decades now, modern physics has proven that *relationships* between these formerly "independent" components are the real key to understanding their functions—that a whole is a great deal more than the sum of its parts. At the same time, academic departments have been moving toward more multidisciplinary considerations, creating schools that include courses combining engineering, history, and anthropology along with biology and ethics. Reductionist thinking is rapidly becoming seen as simply an error of early science. Today we can turn to modern science to prove exactly why the reductionists were unable to predict that acts like owning cars or building factories might have effects outside their social or industrial spheres, that in practice these "isolated" technologies can kill animals and plants in far-off oceans and alter the planet's atmosphere. We're beginning to realize that the inability to predict such outcomes happens when science looks at parts and not wholes.

If we look at a single problem in isolation—for example, energy production—we develop exciting new sources that unfortunately leave uncontrollable poisons behind, such as PCBs or nuclear waste. These "unforeseen" problems are the result of reductionist thinking. To be sustainable, industry and science have to work within the realities of complex natural and social contexts. Today many researchers and even governments are recognizing that developing scientific and industrial technologies that have no reference to their surrounding natural systems and societal values is not only politically or morally questionable. It's becoming dangerous to our very survival.

This kind of scientific progress reflects the general consensus that ethics and cultural values are not "externalities" when it comes to the health of the planet. On a social level, humans seem to need morality, ethics, and spirituality as much as we need jobs and money. Those who practice Second Industrial Revolution principles also recognize that fact and always consider higher human aspirations as they develop ways to feed livestock or build cars. Their long-term goals include these kinds of questions: Will the jobs our new industry is providing be lasting and fulfilling? Is the food farmers grow so wholesome and cruelty-free that they can be proud of it? Do the cities that boast high average incomes distribute their wealth equitably enough to be pleasant to live in?

Let's go over a full list of what's required to survive on this planet over the long term. To be sustainable, any industrial or infrastructure development, any business or management model, whether it manages factories, farms, wild animals, or water resources, has to mimic nature, or it won't fit in with the laws of physics. It has to, above all, be local. It must not produce any wastes that cannot be harmlessly absorbed by natural systems. If it does, it has to reintegrate that waste into the industrial stream.

Like natural systems, sustainable management must be self-regulating, nonhierarchical, cyclic, flexible, diversified—and focused on the long term. Sustainable solutions tend to rise up from the bottom, not to flow down from the top. And like the most creative forms of evolution, they are most adaptable at the margins of security. In other words, poor people in tough circumstances are more likely to come up with long-term, sustainable solutions than highly paid government, corporate, or academic think tanks. Like the examples that lie ahead, such as Allan Savory's efforts to restore degraded grasslands in Africa or the new constitution just adopted in Ecuador, the holistic way is always ready to try something different, and if that doesn't work, to adjust, backtrack and, with care and more humility, try something else. The holistic process toward sustainability is slow because it's always striving for the ultimate goal of complete environmental and social sustainability—and it always keeps its eye on a prize that we may never fully grasp. That's a fundamental principle of sustainability. The people working toward it have to be stubborn in their pursuit of very lofty goals.

IT CAN BE DONE

There are no more technological barriers; there are no more economic barriers; there are only mental and political barriers to 100 percent renewable energy within the next twenty years.
Hermann Scheer[16]

One big reason change is happening so fast is that the writing is on the wall: the age of oil, and therefore much of modern industry and our customary lifestyles, is almost over, if not because of pollution and the reality of global warming, then simply because there are so many of us and so little oil left. New agencies, NGOs, and websites trying to address this overwhelming problem turn up every day.

But despite the widespread recognition of the end of oil and the availability of many new renewable energy technologies, what seems to be going on, especially in North America, is fossil-fueled life as usual. Most of the population seems to be guzzling every imaginable form of conventional energy and is even enthusiastically dredging up the dirtiest fuel ever dreamed of, the tar sands of Alberta. Developing this heavily polluting fuel source shows signs of becoming a Canadian nightmare, as much for its spiraling economic and social costs as for its unprecedented environmental ones. Timid steps intended to address the situation, from the Kyoto Protocol to the very small carbon tax proposed by the Liberal leader in the 2008 federal election, have been derided by government and industry pundits as impossible to administer or too damaging to the economy to even consider. In short, as most people know very well, what stands between viable answers to the global environmental problem and future catastrophe is not technology, wealth, or public interest. It's existing legislation, vested interests, and a lack of political will.

The extremely good news is that some governments have been sufficiently pressured by their citizens to make the hard changes that can bring their countries a viable future. A proponent of using Second Industrial Revolution principles to make economies richer and more secure is Hermann Scheer, a German parliamentarian and economist who's been called "the most influential renewable-energy lawmaker on the planet."[17] He's the author and main mover behind Germany's simple, one-page Renewable Energy Sources Act of 2000, a revolutionary piece of legislation often referred to as "Scheer's Law."

As a Social Democrat and part of the Red-Green coalition that governed Germany from 1998 to 2007, he created a "feed-in tariff" plan to support an ambitiously rapid transition from conventional to renewable energy sources for the country. Under Scheer's Law German electricity distributors must buy power from renewable resources, and they also must pay up to *seven times* market rates for that power. The higher rate varies according to power source and declines gradually over the twenty-year life of the tariff, creating a guaranteed market for renewable energy sources as they become established but, and this is key, *withdrawing subsidies* as renewable energy technologies *prove their economic sustainability*.

Hermann Scheer told *Globe and Mail* columnist Chris Turner in 2008 that his goal is not just to revolutionize energy use. He's working toward "a change of financing structures, a change of the world finance system."[18] If that sounded a little extreme in August 2008, it was echoed by many leaders a few months later when the world's finance system melted down. It has become abundantly clear that "a change of the entire world finance system" is what we're going to need in the twenty-first century.

Scheer's simple piece of legislation is a step along this path. It's more radical than any other green energy law in the world and has added more than 20,000 megawatts of emissions-free energy to the German grid. At the same time, it's generated 24 billion euros in annual revenues. That's *one-third* of the shocking, one-time U.S. bank bailout of 2008 that will haunt Americans for at least a generation— but as an *annual addition* to the German economy, not a subtraction. So Scheer may have a point about revolutionizing finance.

In 2007 alone renewable energy programs in Germany created eight thousand jobs; that means new careers for 250,000 people have popped out of that single page of energy law. These careers will be able to weather the current increase in unemployment across Europe. Scheer's Law has already inspired similar legislation in Spain, Brazil, Portugal, Greece, France, Italy, and California—and is the genesis of British Columbia's, Ontario's, and Quebec's less aggressive policies, which are at least beginning to move in the same direction. Some people complained that using public money to prop up the renewable energy market was unacceptable, even as (usually the same) people demanded bailouts for banks and insisted that car manufacture and fossil fuel use continue to be heavily subsidized. When weighed just against the jobs and revenues that were created, however, Scheer's plan has proven to be not just affordable but richly profitable for Germany. Add the benefits to the planet's atmosphere, and "revolutionary" is the only word for it.

When interviewed for a *Nature of Things* documentary in August 2008, Scheer explained that "conventional energy is essentially commercial"; oil, coal, and gas have to be mined or otherwise dug out of the Earth, transported to be repeatedly refined, then transported again. "But renewables are *noncommercial*. That is, they are limitless and free of cost initially."[19] Every country has wind and sun, as well as heat trapped below its surface. Once we realize that fact, the world of resource haves and have-nots, which necessitates huge, extremely wealthy, centralized corporations to discover, extract, refine, and transport a material good over immense distances—ceases to exist. In Germany right now, many very small villages, individual farmers, or even just householders have invested in biogas, solar roof, or windmill technology for their homes or outbuildings, generating power for themselves and selling the excess to their neighbors and the towns around them. The money they get from feeding energy back into the grid pays off the initial investment, making their own power free within a few years;

it also provides income that allows them to save for their old age or to farm or live in more creative ways.[20] To emphasize his point that any country can have this "indigenous power," which can be put directly into local hands to provide profits even in places not endowed with natural resources, Scheer says, "Although Germany is only 350,000 square kilometers... it's not in a sun belt and doesn't have a big coastline, we were able to create 40 percent of world wind power generation [in only a few years]." How this happened, he points out, is not so much the availability of wind power as it is "the relation between *investment* and alternative power."

Already, 15 percent of Germany's power is alternative; the country's investment in renewables replaces the equivalent production in megawatts of twelve large coal-fired or nuclear power plants. Scheer says, "This is not [so] remarkable... The remarkable data is the speed of introduction"—that is, from almost zero renewables to 15 percent in a little over a decade. What he likes best about the whole scenario is that over the same period in which Germany was implementing this radical legislation, the country was becoming recognized as the economic engine of the most prosperous economic bloc on Earth, the European Union.

Scheer points out that when governments talk to industry about energy sources, conventional energy suppliers aren't going to give objective advice on costs and feasibility, because it's going to be very difficult to switch from being "a seller of oil, gas, coal, or uranium, which are all imported, centralized, commercial materials, to being a seller of wind or solar radiation," which are local, noncommercial, and decentralized free materials. So the conventional energy experts always cite insurmountable technological and cost barriers. "They disseminate disinformation about renewable energy... and many scientists, who also work in the old paradigm, help them."

The situation is complicated by the fact that oil, gas, and coal not only fuel heating, electricity, and transportation networks. Fifteen percent of the fossil fuels mined or drilled are used to produce petrochemicals, from fertilizers and pesticides to wood preservers and nondairy creamers. So if a significant chunk of fossil fuels were replaced by renewables, chemical costs would have to climb dramatically. That creates even more industry resistance.

We're all used to nylon hose, plastics of every description, and daily chemical conveniences like paint solvents and bug repellent. But Scheer knows about renewables there, too. "What does the chemical industry do?" he asks. "They transform fossil hydrocarbons into goods like fertilizers. But plants are also hydrocarbons. What is done already with fossil hydrocarbons could also be done with photosynthetic-produced hydrocarbons: that means plants. And I think the future of biomass is not for fuels for cars. The future of *waste* biomass will be basic materials for the chemical industry. This will make the whole chemical industry much cleaner, much more environmentally sound.

"The problem," Scheer says, "is they always think that these changes would be burdens. They don't recognize that this is a chance for more employment, more prosperity." So far, Scheer's method is winning. In Germany and an increasing number of EU countries, the public and the politicians have gotten the message. Recent charges that the EU is "backing down" from former greenhouse gas targets fail to point out that they're talking about moving from 30 percent reductions of 1990 fossil fuel use to maybe only 20 percent. This is such an amazing achievement that it cannot be considered in the same breath with what is happening in North America.

How did so many politicians in a powerful, industrialized part of the world go green? As usual the answers are both social and historical. Like most European nations, Germany was shocked by the OPEC oil embargo in 1973. Virtually all of its oil was imported, making Germany's economy highly vulnerable to foreign policies. After Chernobyl, Germans were also appalled by the dangers of nuclear power. So back in the late 1980s and early '90s a political party devoted to a sustainable energy future emerged: the Greens. They and a robust anti-nuke, pro-solar public movement influenced the other parties, like Scheer's Social Democrats. Today politicians from these parties have united into powerful coalitions to fight for a solar age as well as for traditional political goals like good schools and responsive health care.

The reason functional democracy is so important to the sustainability of natural systems is that one of the primary requirements of sustainable systems is that they be managed from the bottom up, like nature. All life in the oceans or the savannas rests on humble clouds of phytoplankton and myriad blades of grass. Managing from the bottom up in a human society means that all the people in it—however strong or weak, big or small they are—*must have access to everything they need to survive and thrive*. In other words, to mimic nature, human society has to be as democratic and as equitable as possible.

Parliamentary systems are set up to allow coalitions between parties, rather the way several species in a forest may sometimes compete and other times cooperate. With a majority of the country's votes between them, these coalitions are a balance on any government, permitted to replace it if they feel they cannot work with the single party that got the most votes (but an overall minority) in the last election. Canada has almost never taken advantage of this method of governance, but most European countries do so regularly. The result is that many of them are (most of the time and quite adequately, if their economies and social services are any measure) governed by coalitions of parties, rather than a majority from one party. Voter turnout is twice as high in the EU as it is in either the U.S. or Canada—between 80 and 90 percent on average for most parliamentary governments in Europe. The EU electoral system's responsiveness to the public means that government policies better reflect the public will and also encourages that public to be more enthusiastically involved in policy decisions. This proportional system is far more democratic and therefore more sustainable, which helps explain why Europe is so far ahead of North America in legislation and product management that protect natural systems.[21]

Countless polls prove that most of the electorate in both the U.S. and Canada is vitally interested in more effective environmental policies, that these voters would support the policies even in the face of higher taxes. That means that by taking on more political responsibility and demanding representation more reflective of actual votes, we could quickly get to where Germany already is. And that would lead to some really good news for the environment.

IT AIN'T WHAT YOU GOT; IT'S THE WAY THAT YOU USE IT

In times of economic crisis, the shift to a low-carbon economy is the smartest investment any country can make.

Achim Steiner, executive director of the UNEP[22]

Europe has always led the world in terms of ecological development, largely for the political reasons given above. But today it's breaking away from the pack: not only is it surging ahead in wind power, recycling, and public transport, but the EU has outright banned thousands of proven toxic substances that the U.S. and Canada continue to use. Legal initiatives introduced in Europe in 2001 also imposed energy standards on both individuals and businesses. Today the EU requires a level of recycling North Americans can only dream of. Every car off the assembly line, every piece of electronic equipment, nearly every object manufactured since 2001 has to be *fully* recyclable, usually with *the manufacturer* shouldering the responsibility for accepting the used goods. Old DVD players and cassette tapes, cameras, computers, and outdated iPods don't sit around in your house or leak toxins into landfills. You hand them in when you go back to buy another.

Mark Schapiro, in a 2008 article, "Let's Go Europe," points out that by stubbornly fighting this kind of legislation, U.S. businesses spent many billions more than converting to better practices would ever have cost them. C. Boyden Gray, then U.S. ambassador to the EU, and, as Schapiro points out, "one of the architects of [bank] deregulation in the 1980s," claimed that the chemical and energy laws would "choke economic growth" and "be hell for American multinationals."[23]

Back in 2001, when these laws were being passed in Europe, North America thought it was choosing the safe status quo: a powerful economy over a viable environment. Seven years later, the economic growth rate in Europe had surged a full 20 percent higher than that of the U.S., and, according to the Directorate-General for the Environment, the EU equivalent of the American Environmental Protection Agency (EPA), the only real impact of the reforms has been "to shift resources from polluting sectors to more environmentally friendly" ones. Schapiro says, "studies have concluded that industry, in fighting the new rules, exaggerated the cost of complying; outlays were often half what was predicted." To date, the EU has concluded, there is "*no evidence* that environmental policy has a material effect on the competitiveness of Europe's manufacturing sectors or leads to relocation." Businesses have even admitted they

like regulation, which "levels the playing field and stimulates inno-
vation." Ninety executives of leading companies across the EU say
that the new standards have "ensured both fairness and economies
of scale."

Europe has succeeded in making a profound point about sustainable industries—and U.S. and Canadian manufacturers have
had to comply anyway. North American electronics industries have
found themselves phasing out a large array of toxic metals and
have been forced to make products 80 percent recyclable in order
to get a crack at that 500-million-person EU market. Most of these
changes, according to Michael Kirschner, an engineering consultant to U.S. firms looking to meet EU standards, can be done "in a
non-manpower-intensive manner," since the alternatives have been
available for years. Chemical companies like Dow fought viciously
for decades to keep information about their products away from
consumers. Today, to retain access to the EU market, they've had
to go public with that information after all. Thirty thousand chemicals will be screened in the next decade, and Ethel Forsberg, of the
Swedish Chemicals Agency, says that the net result is that companies are stepping back to reconsider the use of substances they
had, in a knee-jerk way, assumed they couldn't live without. This
proves the ecological economy axiom: proper accounting methods—
that is, internalizing the real costs of substances and technologies
within an industry—will automatically eliminate the use of substances that are too costly to the environment. Forsberg gives the
example of AkzoNobel, one of Europe's largest paint and chemical
companies, which "dropped forty substances from its catalog after
determining that the authorization process wasn't worth the effort."

Eventually Europe's courageous legislation will benefit us all.
Canada has just gotten around to eliminating hormone-mimicking chemicals like phthalates and BPA, albeit only from a few

products like baby bottles. The same action, done several years ago in the EU, has had no appreciable economic effect on manufacturers, although there it's been applied to the entire toy industry. There are thousands of examples of how companies that have been forced to become greener are now outgrossing competitors.[24]

Even in terms of international policies, a new wind is blowing. A few days before Barack Obama was elected president, in the middle of the worst panics about the economic crisis, the UN Environment Programme (UNEP) made history. Under its new head, Achim Steiner, a German economist trained at Oxford and the Harvard School of Business, and as the most powerful environmental agency on the planet, it sought to intervene in what used to be considered none of an environmental organization's business: the economy. Working with the G8+5, one of the most powerful international groups on Earth, and the International Labour Organization (ILO), it launched a "Global Green New Deal" intended to "get global markets back to work [via a] historic investment opportunity for twenty-first-century prosperity and job generation." Norway's environment minister, Erik Solheim, lauded the policy's potential to improve a "very complex situation... a fundamental environmental crisis topped by an international financial situation out of control."

The UNEP is providing its member-states—that is to say, most of the world—with handy guidelines to achieve systemic change, having analyzed with the ILO where most of the future jobs will be found. Pavan Sukhdev of Deutsche Bank says, "Here are... the choices in a nutshell. If we are to lift 2.6 billion people... out of poverty, do we put them into making more and more cars, TVs, and PCs, or do we invest in the protected area network and develop its potential for green and decent jobs?" He points out that a paltry investment of $50 billion a year in subsidizing jobs in 100,000 conservation areas worldwide—that's about a quarter of what Germany is already getting out of the renewable energy industry annually—would not only make sure those natural services are really protected but would generate "millions of new jobs and secure livelihoods for rural and indigenous peoples."[25]

Around the same time, the G8+5, previously a prime instrument for economic globalization, released a mammoth study begun in 2006 on "The Economics of Ecosystems and Biodiversity," referred to as TEEB. Besides making official the "links between ecosystem and biodiversity losses and the persistence of poverty," TEEB demands the internalization of true ecosystem costs by producers as well as rapid implementation of policies for generating green jobs.

The European Union and now the UNEP and G8+5 are not moving toward green jobs and technologies in order to be "good" or even in order to "save the environment." They're looking at the finite nature of some substances, like clean water, and the poisonous nature of others, like petrochemicals and greenhouse gases, and deciding that we must figure out a way to protect some and live without others. As Hermann Scheer explains, not only do such initiatives create a market; they create "a human capacity." The green job pioneers know they can look forward to longer-lasting and more stable careers. "The audience of the professors who give lectures on solar energy, there are hundreds of students," Scheer says, smiling. "And the audience for a nuclear professor, there are five."

As Scheer says with a characteristic chuckle, "Each structural change in economics has winners and losers... We have to make this sincere challenge to all political representatives. 'Which side are you on? On the side of the future of our societies in this race against time, or on the side of the power companies, who say they need... four decades more for their own survival, for the prolongation of their businesses only?' It is very simple."

MAKING THE MOST OF WHAT YOU'VE GOT LEFT

Heavy levels of fossil fuel and particularly petroleum consumption are built into the structure of the present world capitalist economy. The immediate response of the system to the end of easy oil has been, therefore, to turn to a new energy imperialism—a strategy of maximum extraction by any means possible.
John Bellamy Foster[26]

Scheer's is a story about a country with almost no conventional energy available to it. How does the future shake down for a country like Canada, which on the contrary is awash with natural gas and fast rivers for hydropower, and, with the exploitation of the Alberta tar sands and now more in Saskatchewan, has become one of the major oil producers in the world? The 250,000 jobs Scheer's Law has created grow, along with the profits they generate, at a rate of 30 percent a year. There's nothing comparable in energy-rich Canada. If our fossil fuel industry were to proceed at its breakneck pace of 2005—and it had already slowed down considerably by 2008—increases to Canada's GDP over twenty years would possibly be on the order of 9 to at most 20 percent; those increases, however, don't take into account the costs of a horrifically polluted landscape and river system.[27]

This energy is so dirty that even the U.S. Congress under George W. Bush voted to forbid the use of Alberta oil in government contracts.[28] But because the oil and construction giants of the world stand to make a killing, very little effort has been made on the part of the Albertan or federal governments to control spending, distribution of income, or oil royalties. Incomes have risen in Alberta since the tar sands development began, but not in real terms, because housing, food, and transportation costs have ballooned. In fact before the economic downturn, when gas prices were at their height in early 2008, Alberta had the nation's highest percentage of employed people visiting food banks—more than a quarter had jobs. Alberta also has Canada's highest number of poor people per capita, as well as the fastest-rising crime rates.[29]

Deciding to ignore the realities of the New Industrial Revolution, even if Canada does have more fossil energy than other countries, isn't proving to be a wise strategy even for Albertans, much less for Canadians in general. Canada doesn't even have national energy security legislation, which means that in a supply crunch we're not allowed to keep any energy for ourselves but must send it down to the U.S. at the rates we'd agreed to in the North American Free Trade Agreement (NAFTA). Mexico, another oil producer, asked for, and got, an exception for fossil fuels. Canada has never even asked.

Alberta is not only going after the dirtiest energy the planet has ever seen; it's allowing all other sectors of its economy to stagnate. This is termed the "Dutch disease," named after, as a *Globe and Mail* series on the tar sands put it, "the tragic experience of the Netherlands, which discovered oil in the 1970s. As oil exports boomed, the flood of money into the domestic economy inflated the currency, provoked price increases and destroyed exports, leading to a decade of joblessness and rising inequality." Nigeria and Iraq are obvious victims of overenthusiasm for an oil economy, but even Britain did the same thing in the 1980s, when the discovery of North Sea oil "virtually obliterated" its industrial economy, leaving four million people unemployed.[30]

There is some good news, however, even in the realm of the traditional "oil resource curse." Although the Netherlands and Britain succumbed to the general pattern of oil producers, who ignore other sectors of the economy and end up going from boom to bust,

Norway's share of North Sea oil avoided this fate. Throughout its oil boom, Norway managed an annual average economic growth in nonoil exports of more than 7 percent. Alberta established a heritage fund for about 12 percent of the small income it gets from the laughable 1 percent royalty fee levied on the drillers, but Norway's state-owned Statoil and outside firms like Petro-Canada must hand over *78 percent of their profits* to its fund in exchange for the right to drill! In the seventeen years since it was launched, this national savings account "has become one of the four largest investment funds in the world." An illustration of the power of intelligent legislation is a Norwegian law called "the Management Rule" that helps politicians achieve this kind of fiscal discipline. "All but 4 percent of Norway's oil earnings must be placed in the fund for savings." Nothing can be withdrawn until the oil is completely gone, and, "most crucially—absolutely none of the money can be invested inside Norway," which prevents the greedy overheating of the economy that has caused chaos in other oil-producing countries.[31]

This is not only "a profound act of self-discipline," as *Globe and Mail* writer Doug Saunders puts it; it is an almost aboriginal concern for future generations. Most North Americans have barely considered letting their children and grandchildren in for a tiny return on wealth creation from nonrenewable resources. The Norwegians are paying forward the lion's share. "For the Norwegian people," says Yngve Slyngstad, who manages the country's Government Pension Fund, "the oil revenue is not revenue at all; it's just wealth being moved into a more diversified portfolio for the future." In fact, in order to finance social benefits that are among the most generous in the world, Norwegians continue to pay among the highest income taxes, and they shell out $2.30 a liter ($8.70 a gallon) for gas, also one of the highest rates in the world. Despite these facts, they have only 2 percent unemployment and the highest disposable income in the world (owing to those state education, health, and child-care services). And in every quality-of-life index devised, Norway continues to rank near the top.

All of which proves that North Americans need to get a lot more involved in the law- and regulation-making process. Canadians can start by getting more democratic access to Parliament through proportional representation and, yes, coalition governments. Because of oil money flooding Ottawa from taxes, Canada as a whole is in serious danger from the oil curse. And there's another energy threat for Canada and the whole world unfolding. It can be summed up in one question: how can you tell a good energy technology from a bad one?

PAVING WITH GOOD INTENTIONS

Good for the environment and good for farmers. Our government's investment in biofuels is a double win.

Canadian Prime Minister Stephen Harper, announcing a $1.5 billion subsidy plan for ethanol in summer 2007[32]

Producing biofuels today is a crime against humanity... the effect of transforming hundreds of thousands of tons of maize, of wheat, of beans, of palm oil, into... fuel is absolutely catastrophic for hungry people.

Jean Ziegler, UN Special Rapporteur for the Right to Food[33]

When we were writing the first edition of this book in 2002, virtually every other guide to sustainable practices, as well as nearly every environmental NGO, was excited about a new technology to extract "renewable energy" from agriculture, especially corn ethanol. The idea was to *grow* our fuels instead of mining them; oils produced by annual crops like rapeseed, soy, and especially corn would be renewable and less polluting than poisonous fossil fuels. But despite a lot of pressure from friends and colleagues, in our book we didn't advocate crops grown specifically to produce biofuels as a sustainable energy source. That's because agricultural biofuels did not fit comfortably into the Second Industrial Revolution criteria we've outlined in this chapter.

We were then and still are optimistic about certain types of second- and third-generation biofuels like switchgrass and algae, grown under controlled conditions in certain places. But at that moment, when everyone was looking for a magic-button solution, we felt it was safer for the environment and the world food supply to downplay even nonfood crops grown for fuel. Unfortunately, nobody else was applying sustainability criteria to these decisions. Agricultural biofuels were seized upon as a quick, "green" substitute for fossil fuels and massively subsidized by legislation passed in North America and Europe. As the downsides have become apparent, many experts have pointed out that the main reason governments were so quick to adopt food crop biofuels was because they require using so much fossil fuel in the growing process and in processing and transport that their use would barely disrupt the existing fossil fuel industry.[34]

Growing fuels may sound nice, but like every other large-scale management technology, it really depends on who, where, and how. The "first generation" of food crops for fuel, such as corn ethanol and oil palm, doesn't recognize the finite nature of agricultural resources in terms of cropland and water. How limited these resources already are became clear when, only two planting seasons after corn and oil palm subsidies were introduced, Africa began experiencing famines, and food price riots erupted around the world. North America's legislative decision to favor raising corn for fuel instead of food had destabilized prices. EU subsidies on oil palm have caused irreversible planetary damage in Indonesia, where precious virgin forest habitats are being massively cleared to plant oil palm for the European market. Nonfood fuel crops, or "second-generation" biofuels, may prove to be a lot more sustainable, but their use will have to be limited and closely regulated, which will be very difficult in Third World countries. As George Monbiot, best known for his recent book on climate change, *Heat*, said in 2007, these nonagricultural crops, like switchgrass or jatropha, "can grow on poor land and be cultivated by smallholders. But [they] can also grow on fertile land and be cultivated by largeholders. If there is one blindingly obvious fact about biofuel it's that it is not a smallholder crop. It is an internationally traded commodity which travels well and can be stored indefinitely."[35]

This is not to say that certain biofuels couldn't be helpful and sustainable in the short term. On a case-by-case basis, there's probably a place for all of them, although using real wastes for local fuel and only then in properly constructed facilities is undoubtedly the safest. As UN Secretary General Ban Ki-Moon said three years after the subsidy legislation went in, "We need to address these issues in a [more] comprehensive manner." For example, burning food crops like corn or even nonfoods like jatropha or switchgrass immediately puts more pressure on the planet's finite agricultural land base, in terms of bringing more and more marginal land into production. The farmers face pressure to use chemicals, monocultures, and genetically modified seeds to accelerate the production of such crops, which don't have to meet the purity requirements of edible plants.

We know that using corn and soy for fuel means that less of it is raised for food, and as a result the prices on those commodities rise; however, even the prices of other crops such as wheat and barley, as well as milk or even stock animals, immediately shoot up, because less land is available to produce them. Despite what we already knew about fuel pollution and eventual scarcity, millions of people bought suvs in the past decade. *One* fill-up of a 25-gallon tank of one suv with pure ethanol uses over 450 pounds of corn. That's enough food calories to feed an adult human being for an entire year.[36] Multiply that by the number of fill-ups in a week or month and the number of suvs in Canada alone, and we begin to get an idea of why famines and food riots followed so swiftly upon the mass legislation and subsidies that favored agricultural biofuels.

It gets worse. Friends of the Earth Europe estimates that 87 percent of the deforestation in Indonesia from 1985 to 2000 was to make room for palm oil, popular as a source of non-trans-fat oils. With the new subsidies, that already huge rate mushroomed. Loss of habitat for endangered wild species like the world's last orangutans is not the only worry. Deforestation for agriculture also involves draining the peat bogs that lie under the forests. Peat bogs are known to be among the most efficient carbon sinks in the world; when they're drained for oil palm in the tropics, they release huge amounts of greenhouse gases into the atmosphere, eclipsing any benefits using biofuels may have conferred in the first place. Wetland International has revealed that draining peat bogs in Indonesia, some 2 billion tons per year, has now "made Indonesia *the third-largest emitter of greenhouse gases*, right after the U.S. and China, despite the country's small industrial base."[37]

The reason we're including this upsetting story in our litany of good news is to graphically illustrate the fact that even the best intentions can distort markets or create ripple effects on ecosystems that researchers and politicians hadn't anticipated. We have to understand the full implications of ecosystem needs and market pressures when we legislate, and the sustainability rules we're setting out in this chapter will help everyone, worldwide, to do just that. The U.S., Canada, and the EU rushed through legislation between 2005 and 2007 giving tax breaks, mandated use (those "contains 10 percent ethanol" stickers at American gas pumps), and huge subsidies to agricultural biofuels. Such policies, as the Organization for Economic Cooperation and Development (OECD) now admits, have "the unintended consequence of diverting resources from food production" and led to higher food prices, as well as to the destruction of natural habitats. Some of this disastrous legislation is currently being amended, at least in the EU.[38]

So is there any fuel that's safe to use? Roger Samson is the executive director of Canada's most respected think tank on biofuels, Resource Efficient Agricultural Production Canada (REAP), headquartered in Quebec. He advises environmental NGOs like Greenpeace and the David Suzuki Foundation (DSF), political parties, and especially federal and provincial governments on the subject. In February 2008 REAP presented a brief to the House of Commons in the context of federal attempts to amend Bill C-33, the Canadian Environmental Protection Act. The brief laid out incontrovertible data on the costs, agricultural effects, and greenhouse gas mitigation efficiency of biofuels. Basing his opinion on REAP's many worldwide studies, Samson pointed out that introducing massive subsidies for crops into any national farming program, as most Western governments did in the early years of the twenty-first century, "will always destabilize both prices and production." Samson goes on to say that "the kinds of huge subsidies we're using for cellulosic (corn ethanol–type) biofuels are not only destabilizing food prices and harming land bases; they don't deliver on their primary goal: mitigating greenhouse gases." What is needed is *general* farm-support programs, which are becoming the norm in Europe. That way, the farmer can decide what he or she wants to produce, without being forced to grow this crop or that.[39]

Most importantly, REAP's studies confirm recent research showing that corn, soy, canola, or oil palm biofuels either are heavily dependent on fossil fuels for fertilizer, pesticides, transport, and especially processing, or, in the case of oil palm, incite poor countries to convert prime forest habitat to fossil fuel crop monocultures. That means the net result is even *more* greenhouse gas going into the atmosphere than from conventional "clean" fossil fuel sources! Samson explains that we got into this mess primarily because governments "focused on technologies that were commercially ready" and so did not look at alternative crops in detail or even consider a multipronged approach that would have helped out the algae, solar, tide, or geothermal sectors.[40]

Once again, the specter of a finite planet rises up to spoil the dreams inspired by technology and politics. But that doesn't mean we should all abandon thoughts of growing a portion of fuels on our own soil. It's becoming clear that a big chunk of perennial grasslands, the natural form of most Canadian cropland, can be restored by farmers for cash crops that will help heat our houses and move us around, without ripping up desperately needed habitat and watersheds. Samson's studies have found that, along with the judicious use of *used* cooking oils for heating and transport, solid biofuel pellets, from switchgrass or various kinds of grass straw, are the cheapest and most effective ways to reduce the greenhouse gases currently destroying the planet's atmosphere. These grasses can be harvested in the late fall after having served all season as habitat for birds, insects, animals, and soil biota. They need only be cut by the farmer, not planted, plowed, fertilized, or protected from pests, and their transformation into useable fuel requires a processing cost that can be as much as thirty-five times cheaper than other crops. They don't add to the greenhouse gas burden because they demand minimal fossil fuel use, since they're only cut once a year. They require no fertilizers or pesticides, and they keep the land in a relatively natural state conducive to water purification and carbon sequestration.

That's pretty exciting. But before we get carried away and start planting wild grasses all over the place, we have to remember that as part of changeable natural systems and human social responses, even second-generation biofuels will have downsides. If the same crude, sweeping subsidies were granted by governments to switch-grass pellets as to corn and oil palm in 2005–07, farmers worldwide would start producing them on cropland that should be growing food, with the same disastrous effects on food prices. Furthermore, not every morsel of poor-producing oat field, hay field, or pasture should become a monoculture for biofuels. What little wild, marginal land we still have should definitely stay the way it is.

Finding the right balance is a matter of achieving better-researched legislation, backed by the appropriate political will. In the meantime it's urgent that bans or heavy taxes on imports of "bad fuels," including bad biofuels, be put in place immediately, so that countries like Indonesia are not tempted to trash their land and the planet's future with cash crops intended for Western energy markets. The silver lining to the sordid history of biofuels so far is that we can clearly see the enormous power of legislating the *right kind* of pricing. People will have to set very clear goals about actual, not theoretical, greenhouse gas reductions. If these criteria are solidly in place, legislation can do great good instead of enormous harm. We could start using the same mechanisms that enthroned biofuels not just to produce safer ones but to *control the human appetite for them*, the way that EU legislation has controlled the appetites of the public and industry to consume and produce toxins. This means that the systems we use to govern change must be designed with great flexibility. They must be able to respond almost immediately to any evidence demonstrating that a theoretical assumption, like the one we made about agricultural biofuels, is wrong. Which leads to a real million-dollar question: if our very best intentions can go so wrong, how can we guard against further mistakes in policies and technologies—mistakes we simply don't have the time to make?

WELCOME SPACE FOR THE RELATIVES

When I was a little boy, one of my grandmothers had a special thing with moths. At night on her porch, they would come and light on her until her arms and hands were almost covered. I was fascinated and asked her how she did that. She told me she talked to them, and they talked back to her. "What do you talk about?" I asked. "Oh, like if you've been a good boy." I froze, of course. I said, "I *have* been a good boy!" "Well," she said, "that's what they told me."

Henry Lickers, Seneca wetland biologist[41]

Henry Lickers is a Native American ecologist of the Seneca nation and until recently helped run the environment department of the Akwesasne Mohawk Reserve. This reserve bridges the St. Lawrence River, straddling upstate New York and the border area of southern Quebec and eastern Ontario and is composed largely of forest and wetlands. Besides serving on numerous provincial, state, federal, and international boards, Lickers recently won the Royal Institute of Canada's Stanford Fleming Award for outstanding contribution to the understanding of science.

Lickers says that both he and his people, the Six Nations of the Iroquois confederacy, are firmly in the middle of the two most common approaches to both economic and ecosystem management in the modern, Western world. "One side of this debate," he says, "wants to turn animal habitats and nature itself into untouchable environmental museums; they won't eat meat, dairy, or eggs because they think all creatures must be protected from humans, and they believe keeping people away from nature, from exploiting trees or animals or plants in any way, that's the way to help the world."

"Then there's sustainable development groups," Lickers goes on, "like The Natural Step folks, or many of the environmental NGOS, who define what nature is in terms of what it provides for us: the whole idea of environmental services. Nature has no intrinsic value, no right to exist just on its own; it's discussed in terms of human uses, like medicine or energy or whatever."

Lickers explains that, to him, these people are saying that if we humans manage everything just so, we can exploit plants for medicines and farm animals for food, we can control rivers and oceans and indeed all of nature so that it can keep producing everything we want as long as we want it. Exciting techie ideas like geoengineering—injecting CO_2 into the oceans to sequester it, creating biotech drought- or salt-tolerant crops, spraying massive amounts of water into the air to create more cloud cover to mitigate global warming—are at the extreme end of this category. But even conventional wildlife preserves and fisheries laws, which attempt to deal very directly with ecosystem collapse, often assume that all humans need to do to make natural systems work is hone human management skills. "They believe that we can have our cake and eat it too. Instead of not thinking so much about ourselves and trying to step right out of the picture in favor of the animals and trees, like the vegans and preservationists, they see habitat and nature basically as something to exploit, just more carefully. But there's nothing nature has a right to for itself, except insofar as it supports us."

Lickers rolls his eyes good-humoredly. "To us, one side is too far in one direction, the other is off the other way. Native people, I think, are in the middle. Most aboriginals have cultures that think using the benefits provided by the world around us is important to people, but also it's important to nature itself. What I mean is, we're taught to honor a deer as a beautiful, even a sacred thing, but at the same time as a source of food and clothing. So game animals do provide that 'service' the one group is talking about but are also revered as symbol of family and clan, like the vegan group can appreciate. But those last people probably have trouble grasping how you can show respect for the animals when you kill them because you need to eat and use them."

How do you show respect, restraint, and balance in human relations with nature? Lickers says, "For us, that's always been simple: keep the habitat intact, so the plants and animals can live the lives they were given, *where* they're supposed to live them. That's why Indians are always going on about attaining legislation that gives them rights over their part of the Earth. Land claims, that's conserving the land so you can fulfill your responsibilities, not just to the animals and plants and waters, but to the clans you come from— that is, not those animals over there, but the animals you *are*." Lickers thinks one reason animal-based clans are so widespread in the aboriginal world is to divide the tremendous responsibilities such a worldview demands. "I'm Turtle Clan," he says, "My job is to save marshes and rivers. Sure, I know all the science about wetland services; but I get the passion for that habitat because the animals in there are sacred to me and need to be there, in order for my clan, my people, to go on."

Like many other ecologists, Henry Lickers appreciates the fact that "it's easy to understand these connections when people's lifestyles are nested in the environment. So the farmer or fisher is to some extent on the same side as the Native, especially a farmer who wants to stay on that piece of land. Those people understand what we were taught; when you plant apples or corn, expect to lose at least 10 percent of your crop to the animals, including the insects, you share the area with; that's just your due to them. But as people have pulled their lifestyles out of the environment, when they live in concrete downtowns, then the habitat—and our responsibility toward it—becomes unimportant."

Lickers thinks even urban societies can adopt a value system that can work. "A thousand years ago," he points out, "the Iroquois had become pretty urban," Modern archeology supports the surprising fact that large confederacy towns could number as many as twenty thousand people. That was a big city for the time, even in Europe. Lickers says that when many of the great Mexican or South American civilizations became urban, they followed the usual path of chieftains, then kings and emperors: wealth, cities, social stratification, war, and the eventual fall of their civilizations through overexploitation or conflict. But when the Iroquois became more urbanized, they set up a constitution, called "the Great Law," that tried to codify the way of life common to most hunter-gatherer people, in order to hold on to their old values in new circumstances and also to retain a government by equals. "We changed from looking at doing things 'right' and 'wrong' to looking at our actual *responsibilities* to this world and evolved a lifestyle that supported forests, small agriculture, and game habitat, as well as big towns."

Just to the east of Akwesasne, near wetlands on the St. Lawrence River that are protected by UNESCO under the Ramsar Convention, a land claim had been simmering for years. A century ago a huge wetland, part of Mohawk treaty lands, went into the hands of white settlers. In 2008, tired of fighting to have their claim legally respected, the reserve simply bought it all back, bringing an untouched buffer to the east side of the Akwesasne. Henry

Lickers says, "Everyone at Indian Affairs said, 'Why do you want to buy muck and marshland—it's no good for farmland or houses!' The reason is our medicine people say there are many endangered and endemic species there and it's also next to the Ramsar site, so a really viable chunk of this ecosystem can all be protected— 15,000 acres! And to the Quebec wildlife service, we said, we can share: 'Let's work together on a management scheme. Allow us to harvest in there and breathe on our own land, and it can also be appreciated by hundreds of people in Quebec.'"

Lickers points out that "the Maasai, in Tanzania, have been classified by their government as part of the wildlife! In many ways they agree with that; they're part of a huge preserve not just because they've refused to leave but because they're the ones that know how to keep the animals healthy. The Ainu, the traditional people in northern Japan, they have a preserve that saves a crane that would've been wiped out if the government had tried to save it. The Ainu people still kill and eat it, but they also know how to keep its habitat going." Lickers says, "Did you know that over 50 percent of the land set aside as national parks Canada is being preserved because the First Nations peoples *asked* that it be preserved as part of land claims deals? Do Canadians realize that most of that land is in the Arctic and is *de facto* managed by the local people?"

This is good news, because no one has more proven success in managing natural ecosystems sustainably over the long term than traditional peoples. Their record isn't perfect, of course, but with groups that have remained in the same area for very long periods of time, ecosystems are often kept in pristine condition, over centuries and even millennia. Modern, science-based land conservation has caused many catastrophes in only a generation or so—the fundamental error of fire-suppression in forest management, the debacle of overfishing, the steady loss of species in preserves designed around political expediency, not ecosystem needs. So Lickers is one of many traditional managers who have been brought in by government to help. He's been named to serve on both the National Aboriginal Council on Species at Risk (NACOSAR) and the Committee on the Status of Endangered Wildlife in Canada (COSE-WIC), both of which were mandated by Canada's Species at Risk Act of 2007, which officially recognizes "the historical and cultural relationship of many Aboriginal groups to [endangered] species... and the valuable role Aboriginal people can play in the recovery and protection of species at risk."

Although Canadian federal, provincial, and park management decisions can limit the amount of fish or other animals aboriginal groups are allowed to take for their use, the protection situation now works in the opposite direction as well, by making "special reference to the inclusion of Aboriginal Traditional Knowledge in the recovery of species at risk." Because most remaining species are found in native-controlled land (there's a coincidence we'll run into again), the Canadian government has officially recognized "the essential role of Aboriginal peoples in the conservation of wildlife." Both NACOSAR and COSEWIC have mainstream scientists and aboriginal experts as members, recognizing "scientific, community and Aboriginal Traditional Knowledge" as equal components, especially in working out recovery strategies.[42]

An example of government-aboriginal land protection is Quebec's new 770-square-mile Paakumshumwaau-Maatuskaau biodiversity reserve, better known as the Wemindji Project. Located along the southeastern coast of James Bay, about a twenty-four-hour drive northwest of Montreal, it was declared a biosphere reserve by the Quebec government in summer 2008 in order to fulfill the province's internationally agreed-upon UN obligations to preserve 15 percent of its territory by 2010. National and international land preservation agreements are slowly being met by all provinces in Canada and represent another positive example of legislation and political will at work. Wemindji encompasses the watersheds of two large rivers, the Old Factory and the Poplar, which are as wild and pristine as any habitat in all of Canada. The level of protection is higher than in most provincial parks. Low-impact developments, such as eco- or cultural tourism, guided hunting, and fishing, that conform with management guidelines for biodiversity reserves, will be allowed. These latter activities provide opportunities for family-level enterprises among the Wemindji Cree, the new preserve's official managers.

According to hard archeological evidence, the Cree of Quebec have been managing this territory for at least three thousand years, although they've likely been there much longer. About a third of the people in Wemindji make their living in the bush, fishing, hunting, and trapping, and the most respected position in a Cree community is still that of "tallyman," an old Hudson's Bay term for the family member in charge of estimating harvest rates and generally managing a given trapline. An extended family selects one individual as tallyman to manage their trapline, usually a long, narrow territory following the course of a major river. This honor often passes from father to son or son-in-law or from older to younger brother, more rarely to a widow or a sister; but it's very pragmatic. The person selected has to have demonstrated to everyone in the family that he or she is the most gifted—at knowing what's out there, when it's moving, when numbers of various game animals are going up or down—and then figuring to a hairsbreadth how many of each species can be taken without threatening the breeding base.

The tallyman system has worked so well that, thanks to Hudson's Bay Company, which has left fur records dating back more than 350 years, we know that the species mix on Cree territory within Quebec has not appreciably changed in all that time. Such a level of balanced use in what amounts to a desert ecosystem—with animal life largely absent except near riverbanks and estuaries—is a feat that has been matched by a few similar traditional groups, but only in the rare cases where they've been able to manage their territory into modern times. In the new Wemindji reserve, management and protection will take place within the Cree culture's most traditional form. The tallymen of the families with traplines within the preserve will decide how much outside activity in terms of research, hunting, camping, and so forth they will permit. Already, even their staunch and long-time allies, research scientists from McGill University, go nowhere on Cree land if they are not accompanied by that locality's tallyman. They do no research that isn't returned to the community, and the Cree should have the right to refuse the publication of that research.

Although still provisional, this new arrangement is extremely good news. It's a small first step toward adequately protecting migrating species, like nesting birds and caribou or those with huge territories, like polar bears and belugas. The Cree and McGill University researchers know there also has to be a whole system of "cores and corridors" running right across the Arctic and sub-Arctic and down into the rest of the province. Quebec must do what British Columbia, northern California, and the northeastern United States have and enact an entire network of preserves across their northern territories to provide real habitat protection based on the actual needs and behavior of these species. In order to function, such networks must reflect the needs and voluntary assent of the people of the area. Right now in Wemindji, these historic first steps, which both preserve an area and recognize the superiority of traditional knowledge in terms of habitat management, are close to perfect. The right place is being protected, and the right people have been put in charge. In this respect Quebec and Canada can regard themselves as leaders. But they're not the only ones.

MAINSTREAM AT LAST

Nature or *Pacha Mama*, where life is created and carried out, has the right to integral respect concerning its existence, maintenance, and the regeneration of its life cycles, structure, functions, and developmental processes.

Article 71 in the new Ecuadorian constitution[43]

Back in the 1890s the U.S. Supreme Court instituted a staggering legal, social, and environmental precedent that haunts us to this day. It granted legal status "identical to the human person" to private business corporations. Business corporations are clearly not living entities, do not fulfill the scientific requirements of being necessary to human or natural survival, and arguably don't have nearly as much claim on political rights in human society as do the watersheds, animals, soils and biodiversity upon which all life depends. However, their legal "personhood" status has enabled increasingly gargantuan private businesses to overwhelm the needs of not only natural systems but also all humanity. These corporate persons distort the whole "personhood" playing field because corporations don't die like real people and generally have access to almost unlimited amounts of funds with which to defend themselves from other legal "persons"—to say nothing of nonpersons. This is why, prior to the 1890s ruling, corporations had limited powers against real people and could be and were regularly dissolved by local agencies in favor of living beings.

Our irrational recognition of corporations as persons, while simultaneously denying any legal rights to natural systems, has long been recognized as a barrier to any fair or rational legal and legislative defense of the natural world. In *Politics of Nature*, French philosopher Bruno Latour asserts that denying legal status to "nature," in whole or in part, is denying due process to a vital aspect of the scientific reality that surrounds and supports all of human endeavor. In this summary of a long period of philosophical consideration about the assumed "rights" of human beings over everything else, Latour argues for a collective community "based... on a simple extension of the human *and nonhuman*."[44] He points out that until the entire community in which humans survive is granted legal recognition and rights, immediate human needs will always carry more weight in political and economic systems, with disastrous long-term effects for all. Today the old mindset is yet another problem that's at least starting to melt away. Article 71 of the new Ecuadorian constitution, ratified in September 2008, has been hailed around the world as a precedent-setting, legal breakthrough that carries the idea of legal "personhood" one step further. Ecuador's constitution is spearheading a new era in recognizing the legal rights of animals, rivers, trees, and ecosystems as being entirely *equal to those of persons*—theoretically, more important than those of corporations.

Almost a year to the day before the Earth-shaking new Ecuadorian constitution was ratified, the small town of Tamaqua in Schuylkill County, Pennsylvania, also passed a momentous law—a humble sewage sludge ordinance that "recognizes natural communities and ecosystems within the borough as legal persons for the purposes of enforcing civil rights. It also strips corporations that engage in the land application of sludge of their rights to be treated as 'persons' and consequently of their civil rights."[45] This means that the borough or any of its residents "may file a lawsuit *on behalf of an ecosystem* to recover compensatory and punitive damages for any harm done by the land application of sewage sludge. Damages recovered in this way must be paid to the borough and used to restore those ecosystems and natural communities." The fear of liability also tends to put a chill on damaging natural systems in the first place.

Many people will say (and in fact are right now filling blogs with the news) that Article 71 of the new Ecuadorian constitution is just "populist window-dressing" and that elsewhere in the document there are loopholes wide enough for oil companies and developers to drive through with a whole fleet of bulldozers. Corporate interests will be able to continue to despoil the ecosystems that now supposedly have rights. Other critics will also point out, and rightly, that the growing number of small towns in Pennsylvania, Virginia, New York, and New Hampshire that have passed similar "nature's rights" ordinances can look forward to expensive legal attacks, not only from affronted developers and sludge-spreaders but from their own state and national legislators and eventually from international trade regulators like NAFTA and the World Trade Organization (WTO).

The point, however, is that human society is coming full circle. From first seeing the bear and the hyena and even the moth as our relatives, we then adopted a reductionist form of science that separated people from any meaningful connection to or responsibility for them. But this latest revolution in legislation illustrates that even mainstream society, in one of its highest and most influential forms, the national constitution, is again starting to recognize the bonds between all living things and their supporting ecosystems. The judicial system in more than one country is finding the courage to start giving all of nature serious legal, social, and financial rights. Truly modern science is leading the way.

The important thing to realize in all these cases, in terms of the planet's collective future, is not that serious implementation problems lie ahead; the important thing is that *the words have been said*. Eventually people will demand that these words mean something and insist that they become empowered to deliver what they promise. Only a few months after Ecuador's revolutionary words were spoken, in January 2009, Bolivia also ratified a new constitution. The struggle to develop this constitution was very painful. Forces allied with business and the current government's opposition launched a secessionist movement, using gangs and violence, to try to block the national referendum on the constitution that took place in December 2008. Bolivia is the first large country in modern times with an indigenous president, Evo Morales, and an indigenous majority in the population. The "water wars" of 2000–03 were instrumental in the country's changes from a long oligarchy of powerful families to a real democracy based on the actual composition of the population.

The constitution attempts to build a new Bolivia based on the scientific facts of its position: "just enough oil to cause trouble" and a devastated past in which most resources, from silver and tin to bird guano, have been stripped from its citizens by foreign powers. But this is also a region that enjoys the blessings of a large population of politicized indigenous peasants living in ecosystems of unbelievable biological diversity and richness. Among many radical changes, the new constitution nationalizes "all natural resources, renewable and non-renewable," placing them "under the control and ownership of the Bolivian people... the property of Bolivians, for use by Bolivians, for the benefit of Bolivians, and administered by the state"; water access is considered "a human right"; indigenous judicial systems will be elevated "to the same level as existing systems" and non-state organizations such as for-profit corporations will be prohibited from "directly involving themselves in the administration, management, control, and preservation of forests... and natural reserves."

Critics charge that this new vision of Bolivia is economically impractical and will be unfair to non-Indians, "with separate and parallel judicial systems and languages effectively making the indigenous people first-class citizens and everyone else second-class citizens." We imagine all native peoples have some idea what that situation might feel like. But in all seriousness, traditional societies are not easily separated into pieces. To retain their basic identities, they need to have some resort to the tools of traditional justice, language, culture, and values, especially since justice systems are often the primary expression of a society's values.[46]

In *Wild Law*, Cormac Cullinan tells the story of a Kenyan farmer who was brought to justice before his village elders for killing a nursing female hyena that he caught attempting to eat his goats. All parties were represented, with a villager appointed by the elders arguing on behalf of the hyena. The hyena's advocate pointed out that because of a prevailing drought and the hyena's need to feed her young, the animal's behavior was reasonable and the farmer was wrong to have killed her. "The elders then cross-examined the farmer carefully. Did he appreciate, they asked, that such killings were contrary to customary law? Had he considered the hyena's situation and whether or not she had caused harm? Could he not have simply driven her away?" The result of the deliberation was that the farmer's clan had to drive more than a hundred goats—a small fortune in that community—into the bush where they could be eaten by hyenas and other wild animals.[47]

This is an example not only of ecosystem rights but of restorative justice. The community was concerned about its relationships in the larger community in which it exists, which includes hyenas and other wild predators. The hyena's death was taken very seriously, as a general community loss. The farmer was made to suffer for his action and would certainly think twice about doing anything like that again, but all the blame and punishment did not fall on him alone. It was absorbed by the numerous members of his own clan. This not only eases the burden of expense and guilt on the individual but reinforces the idea of society's basic responsibility to educate its members; that is, the clans' duty to train their children in proper relationships—to act with restraint and to show respect to other beings sharing their territory. It also illustrates that small collectives are judged to be just as responsible for such mistakes as an individual.

Any traditional Cree, !Kung, Seneca, Ecuadorian, or Bolivian indigenous person would feel at home with this story, and, to a large degree, with the rights, responsibilities, and type of justice it teaches. But one of the tragedies of the world is that so few traditional people know as much as they should about any of the others. Few North American groups know how many values they share with African or East Indian ones, and vice versa; very few groups know about the new Ecuadorian and Bolivian constitutions. Another poorly understood story that may be vital to human survival right now is the extent to which the traditional and diverse cultural approaches to economics, developed over many centuries, are also being thrown out or forgotten. Thousands of human societies, whether aboriginal, feudal, communal, or mixed, have seen their economic systems replaced by one single choice: a universal belief in that famous ever-growing global economy where everything is a commodity to be bought and sold and new markets and products must continually be found. What is particularly Earth-shaking about the new constitutions of Bolivia and Ecuador is that they have taken the initiative to move in another direction. They are enshrining *diversity*. Even if found in modern documents based on European legal traditions, these constitutions are a first attempt to formally acknowledge the worldview of aboriginal and traditional cultures.

In such cultures other beings, like plants and animals—and even other things, like rocks and rivers—enjoy a level of recognition in human society that is comparable to our own. The belief system underlying the new laws and constitutions appreciates and recognizes *the right of the entire biosphere to exist*. Like Henry Lickers tried to explain, that right is recognized not for philosophical or religious reasons but for what we would term pragmatic political and economic goals. The majority of the population of these countries still believe, as Red Green would put it, that "We're all in this together." If the forests, fish, animals, or rivers are killed, native and traditional peoples viscerally understand it will be impossible for human beings to survive.

Although characterized as "romantic," this attitude could not be more practical. The fact is, the only human social systems that have survived for thousands of years *without* destroying their natural bases are traditional ones. By contrast, in a mere three hundred years, the globalized industrial system has destroyed at least half of the habitat conducive to human survival. The old systems may be uncomfortable and may not let each individual live as long or amass as many goods, but they have never failed in terms of their broad, primary goal: that humans as a group survive and thrive on this particular planet, *along with everything else that supports them*.

Science has proven that a single chemical group, CFCs, intended for the benign purposes of keeping hair attractive and foods cold, is capable of punching enormous holes in our planet's protective atmosphere. It's proven that removing bats from an African rain forest will mean that seeds will no longer be propagated and the entire forest will collapse. It's proven that removing the threat of fires and insect pests from forests will eventually make them much more susceptible to both. Such facts, along with thousands of similar ones, should be teaching everyone that a healthy respect for the interdependence of all the life-forms and resources around us is the safest, most prudent, most economical way to manage our own lives on Earth. Because of political will, this fundamental principle is at last being re-enshrined in modern law. The radically different view of the world that the Bolivian and Ecuadorian legislation represents is a first step toward healthier and more differentiated attitudes about who and what should have legal rights.

The examples given in this chapter each stand in for hundreds more. Hermann Scheer's revolutionary law has made sustainable energy both affordable and effective, first in one country and then across a continent; Norway has scored history's first political and legislative triumph over the curse of oil, through self-denial and a commitment to the future; very powerful institutions like the G8 and the UNEP are introducing global regulatory plans that treat environmental and economic concerns as one and the same; the right kind of habitat and watershed managers are increasingly being given management rights by governments over huge and crucial areas of the planet; and national constitutions are giving long-overdue legal rights to natural systems. These are all are examples of political will and legislative wisdom that meet the strictest criteria of long-term sustainability.

These serious political developments are not covered on the evening news, but the kind of philosophical progress they evince is a real phenomenon. Human beings are engaged in acknowledging that our achievements depend upon the continuing health of all the natural systems of this planet. We're passing the right kind of leg-islation, if we can only learn to recognize it; we even have the first legal precedents in place. Of course, we all need to do more. If we become motivated to shoulder our political responsibilities, we'll have a much better chance of addressing our biggest challenge so far: the impending collapse of so many natural ecosystems. The good news is, if we keep up the kind of work outlined above, the future we build will bring double dividends on every level—from stronger communities to a more beautiful world.

2

USING COYOTES
TO GROW GRASS

BIODIVERSITY

SAVING THE WHALES, THE TURTLES,
AND EVEN THE BUTTERFLIES

My work would not be possible without resorting to the knowledge gathered over centuries by the people of the Amazon.
Juárez Pezzuti, Brazilian ethnobotanist[1]

Hundreds of thousands of dedicated scientific researchers around the world are trying to figure out why a host of living beings, from wildflowers and butterflies to zooplankton and frogs, are slipping into decline. Julian Gutt of the Alfred Wegener Institute for Polar and Marine Research in Germany, who is studying the organisms affected by ocean acidification, and Uzma Khan of the WWF-Pakistan, who specializes in river dolphins and snow leopards, are just two researchers whose studies have recently made news.[2] Beyond simply worrying about vanishing species, however, these scientists are creating programs to save what's left, even figuring out how to reconstitute entire ecosystems. Supporting them in these endeavors are forest guides, hunters, loggers, indigenous people, fishers, boatmen, herbal healers, trackers, farmers, villagers, and many other local people who live where these species still exist.

Experienced scientific field researchers will tell you that they can't function without local help; in fact, if they don't try to get such help, it's probably a good idea to question their results. Rachel Leite, a Brazilian researcher studying turtles and other aquatic life, says that without local turtle hunters or boatman Paulo de Jesus, she would never have been able to find any animals to study in Alter do Chão, eastern Amazonia. "It was hopeless," she says, until her guides explained that the turtles buried themselves deep in the mud during certain tides and seasons and showed her how to dig them out. People who've been on wilderness treks can all tell stories about a guide diving into an innocent-looking bush and coming out holding a rare snake, or a tracker hushing the crowd because he has noticed a forest elephant a mile away. Sights like these remind us that there are a great many skills in life that you don't learn in college. Leite and so many others have learned to just stand there while a local like de Jesus catches half a dozen turtles in pea-green water so thick she and the other experts can make out nothing at all.

"Eurocentric science," says Juárez Pezzuti, Rachel Leite's adviser, "tends to discount local experience. This has hindered research progress and often led to erroneous results."[3] He's far from alone in such a declaration—increasing numbers of academic experts are beginning to share some of the credit, in their scientific papers, with the people who did most of the hard work. More importantly, national and international conservation officials are increasingly turning to local management, in which a multitude of stakeholders—the local users, as well as the scientists, environmentalists, tourists, nearby towns, and so forth—get together to work out comprehensive preservation and use programs for threatened areas of animal and plant habitat.

These new bottom-up management paradigms, known as "living fences," stand in contrast with programs imposed from above by academic or government officials. They use the people who live in a region to help manage and protect it, by honoring their knowledge, considering their needs, and working with them on practical protection methods that will work over the long term. This type of management is being hailed as an exciting new methodology, but it was actually pioneered not far from downtown Manhattan, more than a century and a half ago.

A LITTLE HISTORY

Alewives [are] in such multitudes as is almost incredible, pressing up such shallow waters as will scarce permit them to swim... If I should tell you how some have killed a hundred geese in a week, fifty ducks at a shot, forty teals at another, it may be counted impossible, though nothing more true!

William Wood, New England, 1630

We have hardly any wild animals remaining besides a few small species of no consequence except for their fur.

Timothy Dwight, New England, 1801[4]

One of this book's authors, Holly Dressel, lives on the Quebec–New York border, in a region that's been thickly settled and exploited by humans for over three hundred years. It is characterized by busy, industrialized cities and suburbs, from Montreal and Albany down to the megalopolis of New York. Yet, her home, an old dairy farm just an hour out of downtown Montreal, includes more than the usual farm buildings, woodlands, old pasture, and hay fields. It also supports a family of beavers, two families of muskrats, a pack of coyotes, a herd of deer, a flock of wild turkeys, dozens of nesting ducks and geese, hawks, owls, herons, denning foxes, fishers, and two bobcats, as well as skunks, porcupines, raccoons, dormice, and the occasional visiting bear or wolf. Even a cougar was seen basking by a road only four miles away. The farm contains pockets of very diverse wetland, second- and third-generation forest, wildflowers, reeds, sedges, rare lichens, and many kinds of berries and old medicinal plants. This twenty-first-century miracle of biological diversity has occurred largely because the farm benefits from an overflow of flora and fauna from the Adirondack Park and Forest Preserve, one of humanity's first attempts to legislate the protection of an entire ecosystem and conserve its biodiversity.

The Adirondack Park was created by the extraordinary mechanism of a separate amendment to the New York State constitution. It's only a state, not a national, park, yet it encompasses 6 million acres, which means it's one of the largest preserves on Earth. Although the Adirondack Park comprises only one section of the vast, forested range of the Appalachian mountain chain, it seems to be a big enough section to have made a difference. Even today, although its forests are second- and third-growth in many places, they still shelter nearly all of the vast numbers of tree and plant species they started out with.[5]

Before the first Europeans arrived, this place was one of ecological plenty. The runs of smelt, alewives, sturgeon, and salmon that swarmed up its rivers to spawn were prodigious. Birds of every kind, including the now-extinct passenger pigeon, literally darkened the skies. Deer, bears, squirrels, rabbits, and all the other animals we associate with the Eastern forest were so plentiful that the newcomers started up an economic system using animal pelts, primarily beaver skins, as currency. Yet reports from 1694, barely a generation after the first large waves of European settlers arrived, tell us the deer in the lower section of the Alleghenies, in Massachusetts, were so depleted that the local government enacted a law to close the hunting season. By the 1740s, deer wardens were losing the battle to protect the few animals that were left. And by the end of the eighteenth century, settler Timothy Dwight noted that deer were "scarcely known below the forty-fourth degree of north latitude"—that is, they had disappeared from every part of New England except for the northernmost reaches of Vermont, New Hampshire, and Maine.[6]

The elk, bear, and lynx were already gone; within another generation, by the 1840s, beavers were virtually extirpated, and, even in far upstate New York, raccoons, rabbits, skunks, and deer had also become rare creatures, only occasionally sighted. The reasons are familiar: overhunting and habitat destruction. The new European immigrants converted forests to farms and towns, they hunted and fished wildly, they dammed rivers and diverted and polluted streams, and they started iron-smelting works and timber mills. They cut down trees as fast as they could. By the mid-nineteenth century, not only were nearly all the animals that we now take for granted gone, but their forest was largely gone as well.[7]

The residents of New York State were beginning to learn a lesson—one we seem to have to learn every single generation. The early environmentalist George Perkins Marsh, in his 1864 book *Man and Nature*, publicized the fact that destruction of forests causes major climatic changes and steadily depletes watersheds. And in the early 1870s, New York State employee Verplanck Colvin wrote, "The Adirondack Wilderness contains the springs which are the sources of our principal rivers, and the feeders of the canals. Each summer the water supply for these rivers and canals is lessened, and commerce has suffered."[8] The first "park or timber preserve" Colvin went on to propose was originally a means of conserving timber for later harvest and water for drinking and canals, which were the commercial highways of the day. What ultimately made the vast experiment work were the thousands of park workers, trappers, farmers, residents, and visitors who simply loved the forest and all its plants and animals—loved it enough to fight for it and make sacrifices in order to preserve real wilderness, as much as they possibly could.

It took two long summers of drought and fires, in 1893 and 1894, to build the political will to erect a constitutional barrier that would prevent the timber industry from gaining cutting rights in the park. The final amendment, composed in June 1894, reads: "The lands of the state now owned or hereafter acquired, constituting the forest preserve as now fixed by law, shall be *forever kept as wild forest* lands. They shall not be leased, sold or exchanged, or be taken by any corporation, public or private, nor shall the timber thereon be sold, removed, or destroyed." The amendment was approved by a large majority of the voting population, but what they approved was actually two entities: the Adirondack Forest Preserve of several hundred thousand acres, and the Adirondack Park, a much larger chunk of forest land that was still mostly privately owned.

The state probably had the intention of eventually acquiring all the land, but throughout the 1890s and early twentieth century, rich landowners like William Rockefeller and exclusive clubs like the Adirondack League Club and the Tahawus Club ended up controlling more than 750,000 acres of the park. Since these wealthy and politically influential groups wanted to keep it pristine in any case for "huntin', shootin', and fishin'," the state gradually forgot about the original public dream. Philip Terrie, author of the park history *Contested Terrain,* says that it was "a park like no other park the world had ever seen... It had people living and working in it. It had land owned by individuals, families, clubs, and corporations. It had poor people dwelling in shanties, while down the road were millionaires who summered in mansions... It had land protected as forests forever, and land where cut-and-run loggers savagely exploited the remaining resource in the name of a quick profit."[9]

The park's chaotic mélange of public and private ownership reached another significant boiling point in the 1970s, when the state again found the political will to enlarge the forest preserve from 2.4 million to 6 million acres. Together with environmental groups like The Nature Conservancy, the state bought up many of the old clubs and the family and corporate estates as they came on the market and never lost an opportunity to nab a mountaintop or complete a watershed. Scientists began studies to determine just how much land was necessary to make the park "a viable and lasting entity." With a view to future expansion, legislators enacted more regulations to control the activities of the remaining private landowners. The Private Land Plan and the State Land Master Plan of 1972 limited sprawl and development and made more efforts to curtail logging, protect riparian areas, and reintroduce and strenuously protect wildlife. Terrie describes how Adirondack merchants and town officials were outraged, fearing that the new laws would "violate their property rights and stifle economic potential." State efforts to enlarge the preserve area, these opponents felt, were also "a plot to destroy Adirondack logging and lock up productive lands for wealthy recreationists." Thirty years after the Master Plan was enacted, every move the Adirondack Park Agency makes is still hotly contested. As Terrie says, "The region has become wilder and better protected, and it's simultaneously become more modern, developed, and rife with conflict." Meanwhile, the animal populations swell, the forests look better than ever, and the summer crowds pour in.[10]

A study of the history of the biodiversity protection effort exemplified by the Adirondack experiment reveals many surprising things. As in so many parts of the world—including tropical rain forests, where protection is so desperately needed today—people were living in the Adirondack Park well before conservation was ever considered. Even in the first forest preserve there were towns, villages, and settlements, and the inhabitants weren't just aboriginal groups; they were deed-holding, tax-paying, voting landowners, some of them, like J.P. Morgan, the Rockefellers, and William Seward Webb, very influential indeed. There were also the townies, guides, farmers, remaining Iroquois natives, and local woodsmen who pre-dated them and supplied their needs.

So from the very beginning the Adirondack Park achieved a particular status, precisely because it could not be walled off from human habitation and development, as were other early conservation experiments like Yellowstone and Yosemite that have, from a biodiversity and community standpoint, fared far worse. Moreover, Adirondack forests had been shaved, its streams were often dry, and most of its fauna was gone, just as in much of the world today. Conscious rehabilitation work had to be done. For all these reasons, methods of use and ownership within the park became diversified. The many cabins, farms, and towns inside the park boundaries were subject to a wide and creative variety of restrictions, regulations, and types of ownership. Sometimes people with a homestead or hunting camp in the preserve were granted leases that ran for a hundred years or until their deaths; sometimes they were allowed to keep a place in the family but not sell it. Expansion of towns was limited; landowners are still subject to many land-use regulations that have kept them steaming and complaining to this day. Yet many landowners also supported the park and its conservation efforts so much that they have voluntarily willed their property to it upon their deaths.

Like any other effort to conserve prime land in a heavily populated area, the Adirondack Park has been under continuous pressure to cut trees, make roads, build dams, make ski trails and boat launches, as well as to provide housing and other accommodations for its many visitors. Sometimes it's succumbed to these pressures, and some sections of the park, particularly near Lake George, aren't very parklike. But other areas have also been steadfastly managed for flora and fauna. The Adirondack Park has been an unsung and extremely important pioneer in the whole concept of reintroducing species. This initiative began back in the 1920s with the beaver; the effort to restore this keystone species was an unequivocal success. Since then park managers have brought in many more, from vultures and wild turkeys to moose. Generally, within a very few years of reintroduction, these animals begin to do so well that they start to appear on the outskirts of the park; only five years after turkey vultures were brought back, for instance, they were circling over the farm buildings of southwestern Quebec. Today, for the first time since the early nineteenth century, late summer afternoons not far from Montreal resound with the distant, comical gobble of wild turkey flocks.

As for the rabbits, hares, skunks, fishers, mink, and all the other creatures that had vanished by 1850, they're back in such numbers all over the Eastern forest region that we now take them entirely for granted. We shouldn't. By and large, the wolverines, wolves, and wild cats, and also the cougars and elk, have never really come back—the first two because there hasn't been enough political will to allow their reintroduction. The other big species have remained rare; perhaps the forest is in fact too different, too crowded, or too small to let them thrive.

Beyond the parks, farm and town land all around the Adirondacks, from Vermont and Pennsylvania to Quebec and Ontario, gradually adopted hunting and trapping legislation, which has gone a long way toward giving deer and game birds a breather. But without that huge, prime habitat available to the large animals along the entire spine of the mountains, the American northeast would probably be more like Europe, with some birds, rabbits, and squirrels but no big predators, large flocks of game birds, or wild nuisances like beavers and coyotes. This means that the animals on the Dressel farm, most of which have wandered to the edge of the St. Lawrence valley from that great forest to the south, have literally been rescued from extinction. They are *voluntary fauna* whose important and mysterious lives have been subsidized and saved by humans, and they're out there in great numbers.

If you sit by an Adirondack beaver pond around sunset, the noise from all the animals—the red-winged blackbirds shrilling in the cattails, the swallows twittering excitedly as they scoop up blackflies, the owls hooting from the woods, the frogs croaking in four-part concert, the ducks quacking, the little fish jumping, all soon to be joined by the unearthly yipping of coyotes—simply drowns out any noise from lawn mowers or highways. These creatures are mating and nesting and generally hooting and hollering because we humans decided their lives were valuable enough that we were prepared to give up some of our customary economic activities—not just for a little while, but for a century and a half and counting. They should be a very heartening example that it's possible to balance the needs of a populated area with the preservation of wilderness and biodiversity. The Adirondack Park, for all its problems and shortcomings, is proof that we can save entire ecosystems, with all their waterways, wetlands, plants, and animals, in the midst of large human populations and lots of human activity—if we really want to.

EXPANDING THE MODEL

The "4cs" of Conservation: Core Areas, the main habitats for plants and animals; Corridors, tracts of land that connect the core areas allowing animal movement; Carnivores, to assess habitat health; and Communities, to balance the needs of animals and people.

Adapted from Panda Challenge!
for children on the Smithsonian National Zoo website[11]

Today there's a movement around the world to adopt what some call "the Adirondack Park model" and many others call the "4cs of conservation"—Cores, Corridors, Carnivores, and Communities—which embody a profound new understanding of how to protect biodiversity over the long term and lead directly to sustainable biodiversity solutions on a global scale. The first two cs are clearly explained in the epigraph above. Carnivores like tigers, killer whales, or wolves, as the highest trophic level, are used to assess the health of an ecosystem: if they're gone, it means what's left is too impoverished to support them. "Communities" is the latest and most effective addition to conservation management, as we'll see ahead.

In India, park managers have been paying attention to changing conservation ideas and are much more familiar with the term "Adirondack Park model" than are most Ontario or Pennsylvania ecologists living within a hundred miles of the northeast preserve.

P.N. Unnikrishnan, formerly in charge of tribal protection and eco-development in the state of Kerala, wants to put conservation power right where the Adirondacks put it: in shared amounts between local users and government conservationists. "User-managers are the way to save forests and habitat," he says, and they have economic as well as long-existing cultural motivations for good management. Unnikrishnan is critical of his national government's past policies, saying, "The Indian Forest Service has been doing extensive damage by not using... our tribal communities. They know the value of indigenous forests, especially in terms of medicines. This is an absolute necessity; they couldn't survive without their forest medicines!"[12]

As any good ethnobotanist will tell you, medicinal plants are found in proportion to the species richness of a forest or other biome. The reason we're always hearing about "the race to find cures" in our rapidly disappearing tropical rain forests is that many medicines become rare as soon as the habitat becomes perturbed. Because local users are dependent on the medicines, they will be very aware of actions that deplete the number of medicinal plants. That means if the users are also the managers, they can take some action. "It's not the commercial values of trees the traditional people look for," Unnikrishnan says of India's forests. "It's the variety and utility of species. The state is money oriented, not sensitive to the many other values in life. But people are, especially those directly dependent on the natural ecosystems."

A few years ago, Indian habitat managers were appalled to discover that a national symbol, the Indian tiger, was in serious decline throughout its parks and preserves. The most obvious explanation was habitat loss, but increased sophistication in tracking methods revealed that the traditional means of estimating tiger populations by counting pugmarks was counting the same animal more than once. Poaching is still the biggest problem, however, especially since a single tiger can be worth as much as $100,000. The local poachers are only the bottom rung of a huge, heavily armed network that stretches from local and tribal people across Asia and into powerful organized crime rings in China. It's not hard to imagine poor villagers being tempted, or coerced, into serving this market.

Pramod Krishnan is the director of one of the most famous stretches of wildlife habitat in the world—the Periyar Tiger Reserve in Kerala—which today is increasingly recognized in India as one of its greatest habitat success stories, especially in terms of the future of the beleaguered tigers. Relatively small compared with many Indian preserves, this 300-square-mile sanctuary is home to rare macaques, elephants, three-foot-long giant squirrels, water buffalo, cobras, a small number of tigers, and thousands upon thousands of rare and endemic bird, tree, and insect species. This area of lush forest along the spine of the Western Ghat mountains has long been under pressure for the expansion of nearby spice and cashew plantations. It's also located in a state that has one of the highest population densities in the world as well as widespread poverty. Within that context alone, Periyar has much to teach the rest of humanity about democracy, as well as respecting aboriginal science and sharing the planet with other forms of life. For one thing, its large and very poor population voted overwhelmingly in plebiscites to preserve as forest a *higher proportion of its land mass than any other place in the world.*

Krishnan, a sleek, handsome man in pastel clothes who would not look amiss in a Bollywood musical, has devoted his life to ministering to the elephants and tigers—along with five villages of tribal peoples. Under his direction, the preserve has come to embody the "living fences" of the fourth C, "Communities." The Mannan, Niligri, and other tribal peoples who have lived in this forest for many hundreds of years, and who became poachers when it was designated a national park more than 115 years ago, have now reverted to their original roles as the caretakers of the area. "We protect the tigers," Krishnan says, "because if you can make the habitat viable to them, you're protecting the entire ecosystem," illustrating the importance of "Carnivores," the third C. "But we also protect the villagers, because we've learned the same thing is true." He was one of the first people to realize that in certain circumstances, humans are not just exploiters and destroyers but can be an integral part of wildlife preservation.

"Most preservation systems consider people separate from nature, but in our religious epics, you can see plants, animals, humans—all mixed." Such traditions date back not centuries but millennia. Krishnan adds ruefully, "We used to catch the tribal people when they would come in and steal the animals and the sandalwood trees. We'd beat them up and put them in jail. But when they got out, they'd come back again. This was their home; they didn't have anywhere else to go." As educated government managers like Krishnan tried to keep the area as rich and diverse as possible, they discovered that part of its diversity was due to the sustainable practices of the people who used to live in the area— hard-won knowledge about how this particular ecosystem works, aboriginal science employing the selective clearing, burning, grazing, and harvesting techniques practiced by most indigenous peoples. These techniques all increase "edge"—meadows, clearings, and other variation in forest density that help support more diverse plant and animal life.

So Krishnan did something revolutionary. Starting in 1996, he let the people back in. Not only that, he put them in charge economically. Today, apart from research and entrance fees, *all* the services provided for in the park—the lucrative animal-viewing boat tours of its huge lake, overnight trekking expeditions, village tours, student and youth camps—are managed by the tribal people, and *every cent they generate goes directly to their village cooperatives.* Unlike most comanagement schemes in Canada, Africa, and South America, the tribal groups are dependent not upon changing governments or the policies of centralized managers but only on their own management skills. So their entire livelihood and future is wrapped up in making sure there are plenty of elephants, tigers, squirrels, and trees for the half-million tourists who come to the park every year. This motivation, along with their centuries of knowledge and their profound cultural identification with the forest, has gradually brought tiger numbers back up from an all-time low thirty years ago and has virtually halted the poaching of sandalwood trees, which used to be lost at a rate of three thousand a year.

Tribal guides accompany tourists on day and night eco-treks around the perimeters of the park. These groups unwittingly help the always-armed guides patrol park and even the state borders against poachers. Tourists are kept a long ways away from core wildlife breeding areas but have the impression they're on an exciting deep-forest hike: on a typical trip, the enthralled visitors will see dozens of elephants, scores of wild pigs and wild cattle, the world's largest *and* tiniest deer, to say nothing of several species of macaques and monkeys, fruit bats, moths, and swarms of butterflies. This is animal habitat at its most complex, made more so by the presence of the skilled trackers, who depend on it and take pride in helping others appreciate it.

Periyar has become the bright point in India's extensive history of preservation, increasingly cited as a major exception to some of the mismanagement and failures elsewhere. In its 2005 *Tiger Task Force Report*, a management conference reviewed Periyar's accomplishments with astonishment: "Perhaps India can look after her tigers better by being imaginative in this sphere." The report recommends widespread adoption of Krishnan's innovations across the country. Besides noting the drop in sandalwood cutting and the complete cessation of elephant poaching, the report mentions that, "What the [tribal managers] valued most of all, as they reported to the Task Force, was that they were now treated with respect in their villages."[13]

Already the tribal peoples are more fluent in English than the average scientist coming in from Mumbai or Delhi; Pramod Krishnan made sure they were also trained to identify species, especially birds and butterflies, so that the region's thirty-thousand foreign tourists can benefit from their knowledge and add rare species to their viewing experience. The villagers are allowed to practice sustainable fishing and harvesting methods, especially of medicinal herbs and wild honey. And they continue to share their lives intimately with the animals that make everything about living in the forest possible, as they always have. On a darkening evening a few years ago, two of the local Niligri men showed us around their tiny, thatched-roof village, culminating with a bathing area on its outskirts, a large stone pool in a gorgeous jungle setting marred somewhat by its muddiness. They explained cheerfully, "We can't come here and bathe after this time of day. That's when the elephants need to use it." Indeed, they added good-humoredly, if they tried to bathe when the elephants wanted to, "We'd be killed! So we take our baths early."

FIELD OF DREAMS

If you build it, they will come.

From the movie *Field of Dreams*

One thing we're learning rather late is that it's not just forests full of large predators that need to be understood by the people who manage them. Periyar starts from a cultural advantage, the indigenous and Hindu belief systems in which the needs of animals and those of people are not seen as widely separated in the first place. But the Adirondack Park started as a business-aid proposition, primarily an effort to save a watershed for transportation. It became an experiment in what people thought "wilderness" was in the nineteenth century, and eventually it developed into an effort to save an entire ecosystem. The fact that it happened at all had as much to do with changing values as with increased scientific knowledge in our Western culture.

From the late Middle Ages up until the beginnings of the Romantic movement in the late eighteenth century, Europeans considered trees to be large weeds and forests to be dark, gloomy, and useless wastelands; the only truly beautiful landscapes were the fertile fields and pastures made by humans. The New Yorkers who wanted to save the Adirondacks were at the forefront of a movement that had begun to consider that other types of landscapes might be beautiful and useful but in less obvious ways than a field of rye or a paddock of horses. At first, adherents to the growing wilderness movement were most enthusiastic about forests and craggy mountains, but by the middle of the nineteenth century, people were beginning to realize that without the boggy wetlands they had been filling in or draining for centuries, clean water, as well as birds and other game, would soon be in short supply.

Prairies were also considered to be boring places, blank canvases to fill in with crops or European grazing animals, and, these days, paved-over shopping complexes. But at the turn of the twentieth century, pioneers like the famous American ecologist Aldo Leopold were already trying to restore them. His work on the Curtis Prairie, an old farm owned by the University of Wisconsin, was frustrated when exotics, migrants, tree seedlings, and agricultural weeds all grew merrily in the plowed fields he had carefully sown with native prairie grasses. The unwanted species took over so much that Leopold was forced to acknowledge that his efforts to help nature reclaim its former ecosystem must have been missing a basic component. That component, Leopold eventually realized, was natural wildfires. Fires allow native seeds to sprout, and they get rid of non-natives and saplings. Following a fire, crucial pollinators and seed dispersers are attracted to the rubble. With fire, Leopold's restored prairie has finally flourished, and this flowering has continued from the mid-1930s to the present. Leopold's experiments, chronicled in his famous *Sand County Almanac*, were among the first to trigger the recognition that natural forces like fire are as much the architects of landscapes as geology, seeds, and rainfall.

Since Leopold's day, many remarkable scientists have been working on bringing back biodiversity wherever they live and have let the ecosystem teach them some amazing things. For the past twenty or thirty years, Steve Packard, head of The Nature Conservancy in Illinois, had been trying to restore the prairies of northern Illinois and Indiana. He had mixed success. Although he used controlled burns like the native peoples, the fires did not always remove non-native shrubs; these still had to be hacked back every season so that classic high-grass prairie grass seeds could be planted. Although he was getting poor returns on the grasses, he kept finding other plants he'd never seen before—thistles and wildflowers like cream gentians, yellow pimpernels, savanna blazing stars. Years passed before Packard put the puzzle together and began to realize that the played-out farmers' fields, which he and other botanists had always assumed should be restored to high-grass prairies, had never been prairies in the first place! They had been savannas, "weedy thickets and tall grass growing under occasional clumps of trees."[14] The nineteenth-century pioneers had called what they'd found south of Chicago a "barren." It was neither grassland for their cattle nor forest to be felled; therefore in their eyes, it was completely unproductive. Their "barren," the savanna, is a biome distinct from a prairie and is remarkably dependent on annual natural wildfires. When the farmers suppressed fires, the land had quickly collapsed into thick brushy woods and was forgotten.

Fascinated, Packard bowed to nature and began collecting the "multicolored handfuls of lumpy, oozy glop," as he termed the fruits and seeds of the savanna volunteers, as well as other savanna survivors growing along abandoned railways, old cemeteries, and horse paths. He planted them, and within two years his fields were blazing with species like bigleaf aster, bluestem goldenrod, starry campion, and bottlebrush grass. A drought in 1988 dried up many of the non-native invaders but was just what the reseeded savanna species needed; they took off. Today Packard's fields are full of oval-leaf milkweed, which exists nowhere else in the state, and endangered flowers like the white-fringed orchid, which simply turned up on its own. Bluebirds, missing from the county for decades, arrived to nest, and the classic savanna butterfly, the Edwards' hairstreak, now flutters gaily above the flowers. Packard doesn't know how half the plants got there, let alone the birds and butterflies. Most of the seeds were probably dispersed by animals; but why hadn't that worked before? Others must have lain dormant in the soil for decades, waiting for the right combinations of fire, drought, and companion plants to venture out.[15]

This is a wondrous story, but it also illustrates the confusion we face when trying to restore things we haven't properly identified in the first place. Our difficulty comes partly from the way we've all been taught, since the days of Sir Isaac Newton, to look at parts instead of wholes, individual components instead of interactive relationships. We also view the world almost entirely from within our own cultures. In keeping with the way our society divides up

professions into uncommunicative specialties, prairie researchers like Packard are primarily botanists. Packard began by trying
to restore prairies, which he saw as a community of plants. But of
course, there are many animal life-forms as well. The surprising
return of bluebirds, insects, and other animals Packard had never
expected to the northern Illinois savanna is the natural world's simple statement that the landscapes that we alter intentionally, for our
own purposes, or unintentionally, by introducing exotic species and
climate change, are not just soil, rocks, and plants. They co-evolved
with all the animals—including all the birds, insects, and microscopic organisms that live on them. Each ecosystem—forest, prairie,
peat bog, sedge-covered island—is not a bunch of components but
a whole, a package. *All* the things that live in a place is what biodiversity is all about, and if we don't realize that, we will continue
to cause the same kind of slow, tragic perturbations in natural
systems that occur whenever we try to replace one part of an ecosystem with what we think might work better.

ASSUME WRONG
Relatively high numbers of heavy, herding animals, concentrated and moving as they once did naturally in the presence
of predators, support the health of the very lands we thought
they destroyed.
Allan Savory in Holistic Management[16]

The successful restoration of an ecosystem has a great deal to
do with achieving as much diversity as that system can maintain,
along with having all of its natural components present—fires
and butterflies, turkeys and peat bogs, tigers and people. In the
Adirondacks and Periyar, a major element in the diversity of most

ecosystems—human beings, nature's top predators—were never eliminated; they couldn't be. Perhaps because of its chaotic and undefined nature, the Adirondack Preserve simply aimed at including as many species as it could in its as-wild-as-possible mix of settlement, tourist facility, and forest. But in most government-established preserves and parks in Africa, Europe, and the rest of the U.S., the first thing that managers and rangers did to restore forests and pastureland was to make value judgments about which species they would *permit* to live there. Ranchers across both continents did the same thing, as did the first settlers on the Illinois prairie. They eliminated the indigenous people, suppressed fires, and tried to keep out predators. They also made it a virtue to wipe out as many competing species as possible. They were trying to favor large herds of herbivores and healthy grasslands, and they thought their crops and herds would be more profitable if they didn't have to *share* them with any other species. And yet, all across the American West and on the African plains, in only two or three generations, the grazing animals, whether wild or domesticated, as well as the grass they fed on, began to decline. In fact, desertification is now proceeding across the western U.S. as fast as it is in southern Africa.[17] This fact is teaching us that managing for wildlife diversity is ultimately no different than managing ranches or stockfarms for our own needs, except that we have to look at all the denizens of the land as desirable partners in our endeavor, not just the ones we intend to harvest ourselves.

Allan Savory began his long and varied career as a wildlife biologist in Northern Rhodesia. One of the most puzzling things he noticed as a young man was that after game and livestock had been removed from grassland in an effort to eradicate the tsetse fly, the grassland dramatically deteriorated. Eventually he became a game department manager in Zimbabwe, and in the 1960s his team wanted to reintroduce elephants into grasslands they thought were declining because landowners had overgrazed them. But even with no grazing animals in an area and even with no drought, the grass under their care continued to deteriorate and the land to desertify. After years of disappointment, Savory published a paper reflecting his despair. He concluded that once land was damaged to a certain level, nothing could enable it to recover. A decade later, faced with more catastrophic loss of grasslands in prime elephant country, he made what he calls "the agonizing decision" to cull large numbers of the herd so that the trees and grass could recover. It was a controversial move that made international news, and it certainly wasn't done because the managers didn't love the elephants. He says now, "I thought there were too many elephants. I didn't know how wrong I was until a long time later."

Only when Savory went to West Texas did he began to get a handle on the problem. There he studied desertification, which actually refers to the decline of life-forms in a landscape—that is, the progressive loss of biodiversity. It can occur in wet landscapes, like the Colombian Llanos, but it's usually associated with low rainfall. Like so many other scientists, Savory had been trained in modern, conventional wildlife and agricultural biology, a discipline that evolved in Europe and middle North America. He and other wildlife managers trained in the same system believed that the desertification in sub-Saharan Africa was due to the local people's traditional practices. Scientists and foreign advisers assumed the erosion, plant loss, and soil degradation were being caused by what they observed: too many people raising too much livestock, overcutting the trees, cultivating steep slopes, and using shifting agriculture. These people also used common land tenure, which conventional managers felt worked against their interest in maintaining the land. They were poor and uneducated and had little access to modern research, fertilizers, machinery, or chemicals; furthermore, they were often victims of war and corrupt administrations, as well as prolonged droughts. So when Savory got to Texas, he was amazed to find exactly the same loss of diversity, in the grasses, the insects, and the animals, that he'd been seeing in Africa. The impoverished soil's steady inability to absorb and make use of the sparse rainfall was completely familiar to him from Zimbabwe. And yet, in terms of the tenets of modern land management, West Texas should have been in perfect condition.

West Texas was the exact opposite of South Africa. It had few people, fewer cattle than even a century earlier, and actually too many trees, which were encroaching on the pastureland. The local people were practicing modern, stable agriculture on flat lands that were privately owned; in other words, they had plenty of self-interest in maintaining them. They were relatively rich and well educated and had access to plenty of research and government grants, as well as an abundance of fertilizer, machinery, and chemicals. Furthermore, they had suffered only one year of drought in many, and they lived in peace with a basically uncorrupt government. Yet even with all that, their grass looked as bad as the grass did in Zimbabwe, with more and more ranchers being forced to give up their livelihoods because of the desertification. Faced with this paradox, Savory was forced to conclude that something was wrong with the entire modern land-management paradigm under which he and the Texas land managers had been operating.

Savory took all these conditions into account and devised a new system that classifies land as "brittle" or "non-brittle" on a scale of 1 to 10, with 1 being a tropical rain forest and 10 being a completely barren desert. Conventional management techniques had been developed for what he calls "perennially humid non-brittle environments," like Europe and the best agricultural land in North America, where there is some plant material growing and some dying year-round, some insects and microorganisms remain active at all times, and rainfall is even and fairly dependable. But Africa and West Texas are "seasonally humid, brittle environments"—most of their above-ground plant life dies back for much of the year because of lack of water, and insects and microorganisms also become dormant.

Savory realized that something else had evolved with the grasslands: animals, and especially predators. The lions, hyenas, cheetahs, and dingoes in Africa and the wolves, coyotes, and eagles in the American West had evolved to prey on the huge herds of delicious herbivores that roamed so plentifully on all these grasslands. And then he had his epiphany. He realized that the activity of the predators chasing the herbivores around was as essential for the health of grasslands as the grazing activity of the herbivores—that coyotes and other predators were needed to grow grass. Savory found that in desertified areas devoid of new growth, supposedly lost to regeneration, new seedlings could *only* take root in spots where cattle had been greatly disturbed by a predator. Where the panicked wildebeest or antelopes had bunched together and churned up the ground with their hooves, tearing up and trampling the dead plant material and loosening the soil so that it could receive seedlings and the new rains, regeneration was possible. Moreover, the animals, in their fear, usually remained crowded together for some hours, urinating and defecating. Herbivores will not graze in an area where they have defecated until the material has completely broken down, a process that takes at least a year in temperate Canada. This fact led Savory to his next discovery: the crucial role of timing.

Conventional managers were right in assuming that grasslands needed a rest from grazing in order to recover. This method works quite well in temperate climates like England. What they didn't realize is that in more brittle environments, grasslands behave differently. They don't need an *even* rest over a wide area; they need a period of being torn up and trampled by intensive grazing in small areas, then a short period of rest, then more intensive grazing—exactly what would have happened as they evolved, when the herbivores were being kept in tight herds and chased around the landscape by large predators.

Allan Savory's methods—dubbed "Holistic Management"—have made sense to a lot of people. When we first wrote *Good News for a Change*, almost 2 million acres of ranchlands, particularly in Oregon, Idaho, and Washington State, were being run according to holistic methods, and they are still prospering today. Savory's methods have now spread east, north, and south, to become the desired norm for beef and hog pasturing in most of North America. Even in Quebec and Ontario, farm managers who want to grass-feed their stock (the only feed method that is even marginally sustainable) are told to fence cattle tightly with electric wires and only move the herd when the pasture is really torn up. That way, the pasture will reseed itself with nutritious plants, and the farmer won't have to buy more feed; moreover, the cattle will be healthier. This advice is not attributed by modern farmers to Allan Savory or to any idea of "green" or "holistic" management. It's simply considered modern stock-rearing practice, just as putting animals in open-sided, airy barns is now preferred over the dark, closed stanchions of yesteryear.

In Savory's wonderful how-to book, *Holistic Management*, written with Jody Butterfield, he heads each set of suggestions for land management with two unusual words: "Assume wrong." Savory learned, if nothing else, that natural systems are so complex and interwoven that there is no one-size-fits-all method—even his own. Those ranchers and stock farmers who follow holistic precepts don't do so religiously but adapt them to the needs of their particular area and circumstances. This constant attitude of humility is the most intrinsic part of this management method: because the whole ecosystem one works with is always alive and changing, being extraordinarily flexible and humble in response is the key to success.

ASSEMBLING COMPLEXITY

On the road to extinction, traffic travels both ways.

Kenneth Brower[18]

Doc and Connie Hatfield are part of a remarkably successful western ranching cooperative called Oregon Country Beef. Like many of their neighbors' families, their ancestors arrived in the area generations ago. Once the antelope and elk were hunted out, these settlers started to run cattle. The open range was still home to a lot of other animals: coyotes, wolves, mountain lions, eagles, prairie dogs, and more. The ranchers saw eagles and coyotes take away their calves and lambs and wolves and cougars attacking healthy adult animals; cattle and horses sometimes broke their legs by stepping in prairie dog holes. So the ranchers systematically removed every predator and burrowing animal they saw, by trapping, shooting, and even poisoning them. For quite a while, at least two or three generations, Western beef was king, and the ranches prospered. But by the late twentieth century, the current generation of Hatfields saw their way of life beginning to die.

It's easy to see why, if you drive through eastern Oregon. The area looks like a pale yellow sea, and the only noticeable plant life seems to be one or two species of sparse grass. There's a lot of bare ground, Allan Savory's prime criterion for a "brittle" area. The

Hatfields became interested in Savory's methods in the early 1980s. Doc Hatfield says he had long agonized over "whether it was possible to raise cattle in harmony with the land, and without a lot of expensive inputs like fossil fuel, chemical fertilizers, labor and machinery." But by 1986, the Hatfields were worried not just about the deteriorating condition of their grasslands but also the beef industry's image in the market. Beef was seen as unhealthy, fattening, and full of cholesterol, pesticides, hormones, and antibiotics. The Hatfields weren't using any of those chemicals on their animals, despite the trouble they were having with the health of their grasslands. So they decided to take over their own marketing campaign and began sending about ten sides of beef a week to natural food stores in the area, guaranteeing that it was free of chemicals.

They had trouble getting banks to fund this new venture, because they were using increasingly unorthodox procedures to raise their cattle. They also knew that their main economic problem was wildly fluctuating market prices. So they took another big step: they established an organic beef marketing co-op to set prices based on the average cost of production, plus a decent return on investment, so that members would be assured of a reasonable and sustainable profit. That was nearly two decades ago, and the Hatfields are now in charge of marketing for the Oregon Country Beef brand. It's seen in grocery stores and restaurants throughout the Pacific Northwest, as far north as Seattle and as far south as San Francisco.

Today no fewer than 1.5 million acres of eastern Oregon are holistically managed under the aegis of this one co-op, Oregon Country Beef. The cattle are breeds especially chosen for semi-arid lands, and they're managed according to the co-op's "Grazewell" principles. Chief among these principles is this statement: "Rodents, insects, birds, predators, and other grazing animals all have their role in a healthy ecosystem... Grazings are planned in advance to coordinate livestock presence and forage removal with watershed, wildlife, and human needs."[19] Notice the order of these goals. "Grazing well" is not all about maximizing beef flesh and paper profits. It's about living within the natural world of integrated, interdependent biodiversity, with the belief that cattle will flourish over the long term *only* if other life-forms are around to help them.

Connie Hatfield told us that not only has the holistic method brought them better beef and a more stable market, but it's bringing back wildflowers, medicinal plants, and other prairie grasses that haven't been heard of since the days of the old pioneer diaries. "And," she says in wonder, "in some of our sinks, we're getting water now for most of the year, the first time in living memory!" Did they have to sacrifice their standard of living in order to bring back things like flowers and wolves? "Not at all!" Doc says. "What we've learned is that the economic and ecological are synonymous in the long time frame."

Peter Donovan, a rancher who wrote for a quarterly published by Savory's Center for Holistic Management in New Mexico, pointed out that "This new type of management is something coming not only from the edges of power but *from the edges of economic security*—by that I mean [from] people who are *not* funded by governments, universities, or even NGOs, but farmers, ranchers, fishers, welfare recipients, protesters, and so on. They're driven by necessity to figure out different ways to survive." These people and many like them around the world are evolving similar versions of Holistic Management methodologies without ever hearing of Allan Savory, as they feel their way toward tiger, prairie, and turtle restoration, fisheries, business, manufacturing, or timber-production sustainability. The Holistic Management that we're talking about is just one tendril of an unnamed, growing management movement that works as well with town and even national governance and business management as it does with ranching and protecting habitat.

These holistic ideas seek to steer a different legislative and economic course that will prevent the excesses of infinite growth. Residents of Portland, Oregon, and Freiburg, Germany, are using such principles to help them protect surrounding countryside and to limit their cities' expansion. These cities' holistic visions of their long-term future have no reference to Allan Savory or to one another; they evolved through local culture or as part of the related movement of The Natural Step. In all these holistic systems, however, one thing is clear: any development proposal that doesn't accord with the primary, pre-existing mandate of conserving the entire area—the ecosystem, history, and culture—can be turned down with unusual serenity. Peter Donovan writes that by focusing on long-term goals like community happiness and soil and water balance, instead of short-term problems like a recession or a jump in the crime rate, managers are able to think clearly when presented with temptations, such as too much industrial development or a chance to get rich growing biofuels.

One of the most important qualities used by this unnamed holistic methodology is that it uses biomimicry; that is, it takes its cues from natural, biologically diverse systems. Like them, it is self-regulating, non-hierarchical, cyclic, flexible, humble—and it takes its time. Like Allan Savory trying to restore grasslands, the holistic way is always ready to try something different, and if that doesn't work, to adjust, backtrack and, with care and more humility, try something else again. The holistic process is slow because it's always striving for the ultimate goal of complete environmental and social sustainability—it always keeps its eye on that prize.

ABORIGINAL SCIENCE

Those salmon go all the way out into the ocean. And this is what the older people would think: these salmon, their bodies are sacred. They're gathering all these foods, just as we do. And they're bringing them back to nourish us. The food also nourishes the bear. It nourishes the eagle, the cougar, the animals, the bugs, the microfauna; and that's what the next generation of salmon are going to live off.

Don Sampson, Columbia River Inter-Tribal Fish Commission[20]

Conventional management systems in our modern world are very different from holistic ones. In British Columbia, for example, the trees are managed by the Ministry of Forests; the grizzly bears are the responsibility of the Ministry of Environment; and the salmon are managed by numerous bodies, including some provincial departments and at least three federal departments: Fisheries and Oceans, Indian and Northern Affairs, and Tourism. How well does that compartmentalized, fragmented approach work in terms of the "resources" themselves—the salmon, the forests, and the rivers? Because each department has a perspective defined by its mandate, its budget, its experts, and the bureaucratic turf it defends, the living things that are being managed are never dealt with as biological entities or with a long-term goal in mind. We've spent billions of dollars on government forest and fisheries research without ever acknowledging that the fish, the trees, the water, the insects, the birds, the fungi, and much more are overlapping and interconnected species that are part of a complex and interdependent whole. But these facts have been known for a long time by people approaching the resources from another perspective.

Don Sampson is a member of the Umatilla, a Washington state native group with tribal land at the confluence of two of the greatest rivers on the continent, the Snake and the Columbia. For six years Sampson was head of the Columbia River Inter-Tribal Fish Commission, a group of thirteen native tribes that make their own suggestions about ecosystem management to the many government and industry bodies involved. Like so many other native groups, they've finally gained an audience and a real presence at these official "stakeholder" meetings.

Some years ago, Sampson described to us how certain native leaders had the knowledge and the authority to control where and when people could fish. He gave the example of Chief Tommy Thompson, born in the early 1900s. "He was able to recognize, just by the size and timing of the run and the physical characteristics of different salmon, where they were destined to spawn. So he would know at what specific locations to stop any fishery. The [older people] had a very sophisticated kind of science, therefore, but it also contained spiritual and religious beliefs." Although he's referring to aboriginal science, Sampson understands the conventional kind very well. He's a fish biologist himself, chosen by the elders to be, as he puts it, "an interpreter—because I live in two worlds. I learned the white man's education and I understand their science. At the same time, ever since I was young, I was trained to understand the tribes' philosophy—our beliefs, the principles that we want to manage and live by. I interpret the science from the white man to our leaders, and then in turn I interpret our tribal beliefs to the scientists. But you know," he adds, smiling, "most of the knowledge I gained was not at the University of Idaho, in the Fishery Science Department. It was from the tribal fishers and elders who I worked and grew up with."

Sampson points out that back in 1977 his people were the only ones to realize the salmon were in serious trouble, and they sacrificed their part of the commercial spring chinook fishery to try to help them. Today the commission puts together complex science- and culture-based restoration plans that have gradually won more and more adherents. Their main challenge is to get the U.S. federal government, the National Marine Fisheries Service in particular, to recognize their expertise and to regulate other non-Indian activities so that the salmon can rebuild. Their biggest obstacle is the separation and isolation of all the regulatory bodies, as well as the short-term view our conventional, politically based resource management systems encourage. "Of course the main trouble is that political leaders are only looking ahead three or four years at a time to see how they can get reelected, which is not even the full life-cycle of one fish," Sampson says. "Our plans usually look ahead two hundred years, in twenty-year increments of time."

The tribal vision is fairly easy to understand. Tribe members want to restore the rivers, which as well as being affected by clear-cutting, blocked by dams, and polluted by industry and farming, in some cases don't have enough *water* left in them to support fish. These rivers have been messed up, rerouted, and plundered for agricultural, urban, and industrial needs. And the tribal peoples have had so much trouble getting the various isolated segments of government to see the whole picture—which includes experts trying to figure out why salmon don't do well in rivers that have no water in them—that these days they're working mostly with other stakeholders: the local farmers, industrialists, fishers, and timber barons. "In many cases, the Inter-Tribal Fish Commission actually buys back the rivers, by paying users to return or conserve irrigation water. The tribes are very clear about the fact that private citizens, the other stakeholders in the river basins, are their allies only if their own concerns are addressed. "We're prepared to sit down and work with the local folks, and we know there has to be a gradual transition in terms of their lifestyle, in terms of their needs, before we can all get more sustainable."

"Just a few years ago," Sampson continues, "we put together a vision for the Columbia River, which included all thirteen other tribes in the Inter-Tribal Fish Commission. We asked, 'What do we think this river, this watershed, this home, ought to be like down the road?' And then we laid out specifically how to protect, restore and rebuild the fish and wildlife in the basin: we recognized that health for us humans may come from that as well; not only physical, but economic, health."

This vision led to what Sampson calls "the most recognized success story in the whole Columbia Basin, the Umatilla River Basin Anadromous Fish Habitat Enhancement Project, also known as the Umatilla River Headwaters Plan." The tribes worked with the local communities, the farmers, the ranchers, and the loggers to develop a system where irrigators are provided with a sustainable amount of water from the Columbia. In exchange, they put back into the Umatilla River the same amount of water that they withdraw from the Columbia. The tribes also worked with the Fish and Game and Water Resource Departments to rebuild the steelhead population and reintroduce the fall chinook, coho, and spring chinook.[21]

This reintroduction was extremely controversial, as some of the fish used were from hatcheries and were being allowed to breed, which is considered by purists to be genetically dangerous. "But you know," says Sampson, "immediately the populations began to rebound. And the habitat work that we're moving forward with, augmenting the river by putting the water back in, that's working. The fish are spawning... the nutrients are rebuilding as they die. We're seeing a diversity of species we haven't seen for many years. Bald eagles are coming back to the river... And we're seeing more bears and cougars." Virtually all these animals had been extinct in the area for at least a generation. Forty-five hundred spring chinook returned; less than a dozen had come back prior to the instigation of the Umatilla plan. "And," rejoices Sampson, "in return, we got an economy that is *not* going to hell. The ranchers, they're up there fishing with us now. The local folks are saying, 'We've got all these people coming in. They hear about the fishing, and they're staying in our motels. They're buying gas here. They're buying groceries.' These locals are asking, 'Geez, where are all these people coming from?' And I say, 'Hey, they're coming from the salmon!' "

The salmon has taken its place again in the Umatilla culture. Sampson says, "The greatest thing from the tribe's standpoint, and probably the biggest gift that I ever got, happened when we had our first spring chinook fishery. The young people, and even the generation my father's from, were never able to fish in this river, because the salmon were extinct. And now we see our sons and our grandkids catching them. One of my cousins, she came up to me and says, 'Oh, I'm so happy! My son caught his first fish.' He hooked a salmon; he couldn't believe it, he was so excited. So now he'll be having his first salmon ceremony, where he's recognized as a provider. They'll have a dinner; they'll honor him. He's twelve, and now he knows where his food comes from. He's proud to know that he's a provider, that he has his role in society. It's so great."

GIVING A THOUGHT TO THE FUTURE

Those who think the Aboriginal-nature link is romanticism...
know neither the culture nor the communities.

John Ralston Saul, *A Fair County*[22]

Almost anyone who becomes active in a given environmental issue will eventually find themselves working side-by-side with local indigenous people, if there still are any in the area. There are count-less examples around the world, including the Dayak fighting gold mines in Borneo, the U'wa and Ogoni opposing oil development in Nigeria, the Inuit being witness to the effects of global warming, or the Maya trying to keep their corn landraces pure. These groups are often at the most difficult and violent forefront of environmen-tal issues. Of course there are many places where aboriginal groups have not survived, but even in Italy, Germany, Louisiana, and Nova Scotia, it's the local, long-term managers of fisheries, forests, or farms who end up being the main allies in efforts to mitigate global warming and prevent water contamination or resource loss.

This unavoidable fact has created unprecedented opportunities and challenges in terms of cooperation and especially communica-tion. Gradually, as activists concerned about the environment have absorbed the type of knowledge common to most indigenous peo-ple, a knowledge that is very foreign to Western training, they have found themselves changed and considerably broadened by the expe-rience. And although entrusting the management of a resource to local, traditional users is no iron-clad guarantee for success, it has been proven to work best most of the time. If community mem-bers are strongly identified with an area and have well-established homes there, and if the benefits of the resource are widely spread throughout the user community, then the resources can very often be managed almost indefinitely. It's only when people are allowed to just dip in and out of a resource as opportunists, with no cultural or time commitment to it, or when they have no assurance that tak-ing care of it will bring benefits to themselves or their children, that resources suffer greatly.

This rule goes for aboriginal peoples as well, not all of whom stayed in the same place. And today obviously not every group of traditional people is still attempting to live sustainably or is in a position to prevent sudden resource exploitation. Many traditional societies have lost their language, land, and customs and are in even worse shape, in terms of sustainability, than the larger societies around them. Aboriginal people have often been absorbed into larger societies through enticements, such as jobs, or coercion, such as residential schools. Still, as many of the scientists and researchers who have lived with and studied aboriginal groups testify, it's not a coincidence that most of the remaining relatively pristine waterways and habitats on Earth are in aboriginal territories or are inhabited by very traditional groups. Sometimes this is the case because governments sought to protect these refuges for such peoples, but more often it's because these reserves and preserves are where the traditional groups were originally from and insisted on living. Today in Canada members of aboriginal societies are free to leave their tribal areas for cities and suburbs, but a significant number—46 percent, close to half country-wide—choose to remain in their home territories, and many who do leave go back. This phenomenon is global.[23]

The many aboriginal people remaining in or returning to home villages are choosing to lead what may seem remarkably restricted and difficult lives. They will often say they are trying to remain involved in practices characterized by words in their languages that mean "restraint," "respect," and "human responsibility." Some of these words imply constraint or difficulty in English, but that's an imperfect translation. In fact the type of life most aboriginal groups try to live focuses on long-term survival within the physical realities of a locality. The idea that human beings are independent from the land or the water around them is foreign to most aboriginal groups. And today, physics and evolutionary science have proven that the idea that humans are separate from nature is a fantasy. More proof of its unsuitability to this planet is the fact that this idea has led to technologies and belief systems that work very hard to ignore physical reality and have rapidly destroyed the habitat of most of life on the planet.

Laurie Suitor has been working with aboriginal groups so long she uses the word "we" when she describes native belief systems. She's the Intergovernmental Relations and Partnership Advisor for the Unama'ki Institute of Natural Resources, near Bras d'Or Lake, which is managed by the Mi'kmaq First Nation in Nova Scotia. Suitor explains that "there are two kinds of science: Western science and what is now widely being termed 'aboriginal science.' The latter differs in methodology... Instead of observing dead specimens in a laboratory for the brief period of an experiment, aboriginal science observes the living thing, in a living state, in a living environment. And over a very long time; that is, many human lifetimes."

Fikret Berkes, an anthropologist at the University of Manitoba with many years of working with indigenous groups under his belt, would concur with this simple definition. While explaining that many indigenous languages don't have words for "management," "resource," or even "nature," he points out that this fact hasn't prevented them from developing intentional effects on the things these words describe. Speaking of the local fishers working on the rivers flowing into James Bay near Chisasibi, Quebec, Berkes says, "The Cree do not have formal management policies, but they certainly have customary practices, that, like the policies of management agencies, can change dramatically." Berkes likens their practices to a term more familiar to scientists: adaptive management. "The Cree assume that they cannot control nature or predict yields; they are managing the unknown... an unpredictable, ever-changing environment, and [they] are experts in using resources at different scales of space and time... the Cree hunter-fisher has respect for complexity and uses practices that conserve ecosystem resilience." Looking for a bridge between the two management cultures, Berkes says the term "traditional management" can be used interchangeably with the scholarly idea of "adaptive management."[24]

Darrell Morris, the Mi'kmaq project manager at the Eskasoni Fish and Wildlife Commission, says that indigenous ecosystem managers have very specific goals. "The purpose of studying these living creatures and living systems is not the Western ones of curiosity, industry, or use: it's for learning how to live. We learn our moral and social philosophy from the natural world." Laurie Suitor gives the example of traditional houses—portable wigwams and temporary bark longhouses. While aboriginals in Eastern Canada had the numbers, strength, skills, and materials to build permanent wood or stone dwellings, their science—their observation of regular, periodic crashes of animal populations in the Canadian ecosystem—had taught them not to invest too much in staying in one place. "They learned: don't leave a footprint; don't get too big; use natural materials."

Suitor acknowledges that "Western science gives you really interesting parts of the truth; the substance in this experimental action leads to this reaction or product in this particular laboratory setting. But you get tragedies in the real world when you take laboratory science out into it." Society keeps experiencing these science-based tragedies, from thalidomide and bisphenol plastics to nuclear power, DDT, and now biofuels. That's why Suitor thinks everyone could benefit from adding aboriginal science to mainstream methodologies, so that people would have "teachings which inform your values, which in turn *regulate your behavior*. So you fit in with the natural world and don't hurt it."

Suitor and Morris's group has recently raised about $240,000 to preserve their area's environment, animal habitat, and social customs and to build a traditional village on the Bras d'Or, the beautiful inland sea alive with fish and eagles, of central Cape Breton Island. "We're working with municipal planners to get nature into the mix," Suitor adds, but she worries that "there really is no recognition or appreciation of Mi'kmaq conservation, of aboriginal science," either in Nova Scotia, where they have to try to get funding, or in Canada as a whole. "*Context* is what we provide to the limited, isolated ideas that come out of mainstream science," she says. "Aboriginal science rests on three principles: (1) Humans are dependent. We are not in control of this planet, we are not in a hierarchy to affect management so that it benefits us, either exclusively or primarily. (2) Human beings have to understand that we don't manage ecosystems; they manage us. (3) Our main job is not organizing things for ourselves but for future generations."[25]

What is really key for the rest of us to learn, Suitor explains, is the fundamental principle upon which this other form of science rests: "Human *restraint*, not management, is what we need." The history of most aboriginal groups arguably can be looked upon as a history of restraint. Not taking too much, not having too many children, not leaving too much behind. Aboriginal groups have long interested population biologists because of their ability to keep their populations almost steady when they are in their natural state. The Cree of Quebec, for example, had a population that hovered around five thousand people since at least the 1600s and only began to double and triple around 1970, when Quebec and federal policies took their lands and traditional food base and forced them to settle in towns or flew them to residential schools.

Morris says that when he attends various environmental and scientific conferences, "Non-Indians are always saying we need more maps, graphs, statistics, or computer models, in order to figure out how to manage whatever it is we're concerned about: a river, the otter population, forest cover. But our elders say we need to *stop our activities.*" "As a matter of fact," Laurie Suitor interposes, "we've amply proven that humans cannot manage ecosystems. What we can manage, if we'd try and look at all these groups who do, is not nature, but *ourselves.*"

This makes sense in terms of the incontrovertible fact that when humans are successful at "managing nature," for example in maintaining the Adirondack Park, we are arguably not managing nature but ourselves. When we're lucky, nature, in its generosity, can flourish despite our interventions and restore its abundance. When it fails, the first thing to try is self-control. Laws that constrain development, housing, and other perturbations in wetlands and forests help us manage ourselves so that the wetlands and forests can get on with their own activities. This traditional or adaptive management approach, as opposed to intrusive management—like preventing all fires or removing predators—is being increasingly adopted today, because so far it's the only one that's been demonstrated to deliver healthy ecosystems.

LIVING FENCES

One major reform principle underlies all the specific and general [habitat preservation] reform ideas: Involve the community.
Wende Pearson and Ryan D. Andrews[26]

The land is the boss and will teach whoever she wants.
Rene Lamothe, speaking of the Dene[27]

Few people realize that when Yellowstone Park was established in 1916, it wasn't empty of humans. Various Native American groups continued to live and hunt in it during the first years of its existence, but gradually the late-eighteenth-century European idea that "nature" does not include "humans," then considered a special outsider group closer to angels than animals, took precedence, and they were forced out. Professional park managers, many of them off in Washington and most of them strangers to that ecosystem, did not realize for nearly a century that the native practice of controlled burning helped produce the rich wildlife and plant species that made Yellowstone worth preserving in the first place. By the 1980s, centralized policies of repressing all fires and eliminating certain animals like timber wolves led to tinderbox forest conditions that caused the worst fires in history and allowed many grazing species, like elk, to massively overpopulate and overgraze.

Today, Yellowstone's managers are playing catch-up, doing controlled burns, reintroducing wolves, trying to mimic some of the management practices of the people they kicked out. Elsewhere in the world, this management approach is becoming part of modern preservation legislation, and the real experts are not just being imitated but are being consulted and given real power over the land. Of course none of this new legislation has come easily. There have been decades of jailings, beatings, landgrabs, overhunting, and overfishing, along with well-intended but tragic mismanagement of wildlife; in short, general waste caused by ignoring aboriginal science. But where native people have managed to hang on to or reassert their rights, things are beginning to change.

The Amboseli National Park in Kenya, which was always used by the Maasai people for grazing and plant collection, is now largely managed by them. Under government protection by 1906, a full decade before Yellowstone, Amboseli was originally a rare example of a multiple-use system that granted the Maasai full land rights. This system worked remarkably well for fifty years, until that philosophy of nature versus people took hold following World War II. In order to achieve the government goal of "preserving" wildlife, which at that point hadn't been appreciably depleted, the Maasai were deprived of their lands. More changes detrimental to both Maasai rights and the park's biodiversity were implemented in the 1970s. Bans on hunting and cropping created an overpopulation of certain species that led to more overgrazing and soil erosion both in the park and around its borders. Tourism was eating away at Maasai culture, introducing alcohol and drugs, causing "destruction of traditional values and respect" within the community.[28] Early plans to give control back to the people were rejected, and in retaliation the Maasai "killed large numbers of wild animals to exhibit their disapproval."[29]

The government was forced to negotiate, as it often is when local users resort to sabotage and noncooperation to force an issue. Eventually the World Bank, in exchange for funding a long-delayed water pipeline for the Maasai, mandated a compromise that included a grazing compensation fee, permission to hunt and grow crops on some land, and accommodation of tourist campsites and lodges.[30] Since these concessions were made, employment in the park and control of tourism have provided the Maasai with incomes far higher than merely grazing their animals had and has created a new bond between the park and its ancient users. More refined investment policies and mutual management initiatives have followed, making Amboseli today "one of the most cited examples of protected areas returning benefits to local communities."[31] The first $30,000 compensation fee was used to build a community school, an obvious benefit to the Maasai, and, to the government's surprise, "the number of wildlife in the Amboseli ecosystem has increased significantly. Elephant and rhino numbers are rising because of reduced poaching. Zebra and wildebeest populations have also increased because of less competition with domestic stock." Even more excitingly, "Wildlife populations are also exhibiting more natural distribution patterns," which indicates long-term viability.[32]

Only now are Western-trained researchers realizing how important the Maasai have been in terms of maintaining that biodiversity. Recent studies have discovered that the Serengeti grasslands are recovering in part *due* to the grazing and migration of Maasai cattle, not in spite of it, as previously thought. "The cattle prevent scrubs and woodland species from overgrowing, which allows for more grazing opportunities for animals such as antelope." Allowing the Maasai to live traditionally in their own territory is illuminating a human-management-grazing effect. It mirrors the North American indigenous practices of controlled burning, along with Allan Savory's discoveries about grassland decline in Zimbabwe.[33]

A similar success story is unfolding in Canada, where the first national park to be comanaged by a native group, Gwaii Haanas, had a birth no easier than Amboseli's. Even though the Haida people never signed treaties ceding legal authority over Haida Gwaii—what used to be called the Queen Charlotte Islands off the north coast of Vancouver Island—all the islands of Haida territory were opened from the nineteenth century onward to large logging companies and commercial fishers. Decades of habitat destruction resulted in the famous Haida blockades against logging in the 1980s and led to a key development. As one assessment put it, "The alliances formed at Haida Gwaii between First Nations and environmentalists during the 1980s helped to change the way environmental activism was conducted in Canada."[34] By the early 1990s the federal, not provincial, government designated the new park, partially in recognition of the fact that one island of this stunning archipelago had been named a UNESCO World Heritage Site ten years earlier. The Haida nation then obtained a comanagement agreement for the park islands to spare them from further logging. But because massive logging continued on the rest of the territory, the Haida also filed an aboriginal title lawsuit, resulting in a 2004 Supreme Court decision declaring that provincial governments must consult with the Haida regarding any development. This ruling followed the landmark 1997 Delgamuukw decision, in which the Canadian Supreme Court prodded provincial governments to negotiate treaties with all aboriginal groups and establish who has jurisdiction over what lands.[35]

Today these decisions remain insufficiently enforced, and the Haida, like many other native peoples, still have to resort to blockades and seizures to get media attention concerning their treaty and land tenure situations. The Haida have been particularly effective because they have allied themselves with environmentalists and forced legal recognition in the highest courts in the land. Significantly, they work closely with fishers, loggers, and nearby municipalities to maintain the long-term viability of this once incredibly productive ecosystem. Together they're fighting offshore oil drilling and oil tankers; they're opposing the recreational hunting of a rare black bear subspecies that nets private businesses between $7,000 and $11,000 a bear. And they're guided in these decisions by aboriginal science. Haida Council President Guujaaw, a very old friend of David's, explains how difficult decisions, such as opposing the bear hunt, are reached. "Our people have taken a position that it is not an acceptable form of economy for our islands. Killing bears for fun does not meet the moral tests of proper behaviour for human beings."[36] In other words, although these bears may be hunted for food or clothing, trophy hunting does not demonstrate the restraint, responsibility, and respect demanded by aboriginal science.

The latest comanagement agreement, of December 2007, gives the Haida nation rights over 53 percent of Haida Gwaii. This is another legal precedent, because it means British Columbia has finally recognized the Haida as a governing body, not just an "interest group." Guujaaw told the press, "This agreement is the first land-use plan on the islands, aside from the 'divide the loot' kind of approach that we have seen over the last century."[37]

The Maasai and Haida stories are examples of wise court decisions and intelligent legislation at work. Both are based on the concept of paying attention to what has worked in the recent past as well as to the intricate relationships inherent in any natural system. Most importantly, it means remembering that those systems almost always include humans. Like Allan Savory's major concept of "assume wrong," one Amboseli study proclaims that *all management is a long-term experiment,* and decisions are always made with less than complete information."[38] In other words, only adaptive or traditional management strategies work in the real world of nature. These stories also illustrate one of the primary laws of sustainable systems: answers come from the bottom up—from the least mainstream sources—not from the top down. Solutions often come from people living in these margins of security, because they are in a position of humility and are ready to adapt to new conditions. In these contexts the Maasai and the Haida are perfect examples of sustainable managers. They have to survive where they are because they have nowhere else to go and no other way to live. That means they have to *listen to the land* and develop ways to survive that not only do not deplete other life-forms but help them flourish. The results of their many generations of difficult and attentive research, in both cases, speak for themselves.

GOING NATIVE

When do we all become indigenous people? When do we become native to this place? When do we decide we are not leaving?
Bill McDonough, architect and designer[39]

In 1994, in the wake of the first Earth Summit in Rio, the elders of the Mohawk nation wrote their own environmental restoration strategy. This document was presented to the UN in 1995 and is enjoying new attention. It asked for the involvement of native decision makers "on a government-to-government basis in all discussions regarding risk," because the Mohawk saw themselves as experts on the understanding of survival. Their primary ecosystem restoration plan, however, was simple: "Take no more than you need."

That simple concept of restraint may seem unimaginable to beings as spoiled as most of us have become; do we *need* BlackBerrys and the latest iPhone, a new outfit stitched up in Sri Lanka or battery-operated robotic toys for our kids? Who decides what we need—Madison Avenue, the WTO, the federal government, our local Chamber of Commerce, we ourselves? The famous green architect and designer Bill McDonough has been trying to make this point for a long time: Earth is the native home of all humans. All of us, however Western and globalized we have become—and whether we admit it or not—are natives of this place. Without exception, the ancestors of every human being now on Earth once lived in small, hunter-gatherer or horticultural groups that had to understand how to husband the Earth's resources and how to limit their own demands and desires. Even as those forebears refined agriculture and developed more sophisticated technology, most of them continued to hold on to basic "green" ideas: the positive value of thrift, the reuse of all materials, the importance of sharing resources and not boasting about good fortune. A mere sixty years ago in Canada and the U.S., being thrifty, "making do," helping neighbors with no thought of repayment and considering it only good manners to avoid ostentation were common and highly praised virtues among populations fully as happy—in fact, according to polls, demonstrably more so—as we are today.

All this means that we could become "native to this place" again rather quickly, as thousands of initiatives around the world demonstrate. For instance, the Maritimes is one of Canada's poorest and most industry-dependent regions, with a long tradition of corrupt and unresponsive governments. However, a new law was passed in Nova Scotia in spring 2007 that sets a remarkable precedent: the Environmental Goals and Sustainable Prosperity Act (EGSPA). Unlike many discussions using the S-word, this one embodies a careful consideration of reality, backed up with teeth. It makes Nova Scotia, in the words of the province's minister of environment, Mark Parent, "the first province in Canada to legislate a broad range of environmental targets *with the explicit intention of fostering sustainable economic prosperity.*" There are those two ideas together again: the economy and the environment, indissolubly linked by legislation mandating immediate, enforceable, specific environmental targets.

Between 2000 and 2005 the province of Nova Scotia set itself a seemingly impossible goal of cutting its production of solid waste—garbage, that is—in half. When it achieved its ambition with time to spare, the experience opened up minds the same way Herman Scheer's ambitious law did in Germany. Now Nova Scotia has as an even loftier goal: that the province will have "one of the cleanest and most sustainable environments in the world by 2020," and that by the same year, its economic performance will be "equal to or above the Canadian average," something that also has not been the norm for the Maritimes. Critics say EGSPA goals for greenhouse gas emissions, forest health, coastal protection, and sustainable transportation could be considerably more ambitious, although significantly they remain flexible and can be updated. The main point is that here is a Canadian province that has serious green economic goals and the legislative means to accomplish them.

Some of the better goals include a mandate that 18.5 percent of electricity needs will be obtained from renewable sources by 2013, that California emission standards for vehicles will be in place as soon as 2010, and that 2001-level sulfur dioxide emissions will be halved by 2010. These hard targets force many different government departments to work together, already a sea change in philosophy, to produce a single document accessible to the average citizen, whose inclusion is more sought-after than usual. Many NGOs and other stakeholders, including GPIAtlantic's Ron Colman, were given input—for example, to change wording like "economic growth" to "economic development." Colman's group works to measure societal well-being through a "General Progress Index" that seeks to value such things as schools, environment, and equality, not just cash transactions. He says the new act, "suggests there is value in qualitative changes that reflect well-being—this is basically Herman Daly's idea that bigger is not necessarily better."[40]

An example of new ecological and economic stewardship almost immediately emerged a short time after this legislation was passed. Bilcon, a Delaware-based concrete company, wanted to build a huge rock quarry on Digby Neck, a spit of land jutting into the Bay of Fundy. Pollution, noise, and shipping were seen to be major threats to the endangered right whale population and the area's thriving lobster fishery. Moreover, the community valued its position off the beaten track in one of the most beautiful parts of the Maritimes. Even though the area is economically depressed, the residents rejected the possibility of thirty-four new jobs or the idea that any kind of "development" that endangers existing resources is desirable. To everyone's shock, at the end of the formal environmental assessment process, both the federal and provincial governments agreed with the locals. In Canada thus far, environmental assessments have never completely stopped any industrial development, however damaging; they have sought only to mitigate the worst effects or to limit or contain the worst practices. But in this case, the assessors outright canceled the quarry project, writing that "It would have a significant adverse effect on... the 'core values' advocated by the communities along Digby Neck and Islands... [leading to] irrevocable and undesired changes of quality of life."

The use of the terms "core values" and "quality of life" as the key objections to a development is unheard of in such documents and has to reflect some of Nova Scotia's new sustainability promises. But as always, the real impetus came from below, not above. Hundreds of residents had come to the hearings to try to express their spiritual connection to this thin spit of land. In her report to the assessment panel, government consultant Janet Larkman wrote, "My encounters with the people of Digby Neck have taught me more than any textbook ever could about what it means to love your community, and to love the land and sea that nurtures that community. They have articulated with unrivaled elegance what their community means to them, and have fought vigorously against forces that seek to tear them apart."[41]

Bilcon is contesting the decision with a challenge under the infamous Chapter 11 of NAFTA, a skewed trade agreement that grants corporations more rights than democratically elected governments. They are unlikely to succeed; it would be politically inadvisable for the Canadian or Nova Scotia governments to change their position, given the popularity this decision has earned them. They are also unlikely to do so because the panel that assessed the project went far beyond just canceling the Bilcon development. In another unprecedented statement, the panel called for "a comprehensive coastal zone management plan [from the province], a moratorium on new quarry developments [in the area], better mechanisms for consultation with communities and local governments and environmental assessments for quarries of all sizes in Nova Scotia (not just ones larger than 3.9 hectares [9.6 acres])."[42]

The lessons of "going native," which are based on looking at how the limits of the physical reality of this planet teach humans how to behave, are expanding all over the world and at every level. They lag behind in some areas, like our treatment of the world's oceans and of its greatest gift, fresh water—but they are steaming ahead in others, like forestry and agriculture. In every case, there are clear paths to follow as well as clear criteria that can—and must—be met. The goal is to be able, as another native group, the Mohawk traditionalists, says, to provide for the needs of all those that humanity is responsible for: the planet's unborn children of our own and of all species: "the faces yet beneath the Earth."

AVOIDING VENUS

CLIMATE CHANGE

FIXING THE CLIMATE

What we need is comparable to the mobilization for World War II. We are talking about radical, fundamental changes... People have to realize the enormity and magnitude of the effort that is going to be needed to turn this around.

Danny Harvey, climate specialist and Nobel Prize laureate[1]

Ranks of huge, propeller-topped towers marching across the horizon, feeding wind energy into power lines; more mysterious towers dotting the rural landscape, growing algae for biofuels; painted-on solar coatings that turn almost any surface into an energy-producing solar panel; pumps that warm buildings directly from the Earth's molten core; electric cars that feed energy back into the grid. All over the world, visions of a high-tech, green future are firing people's imaginations. Green technology is even seen as a way to save the global economy from its current spiral, now that the casino bubbles of dot-coms, housing, and finance have burst. Exciting concepts to be sure, but how many will really work? That is, how well will these technologies achieve the real goal: reducing the carbon load in the planet's atmosphere? Do any of these solutions hide unexpected inefficiencies, social or ecological disasters, or special-interest economic agendas? For example, are the space-age technologies used for certain kinds of algal biofuels or thin-solar production safe and sustainable over the long term?

As well as answering such fundamental questions, anyone seriously trying to address climate change will have to check whether high-tech solutions can come online quickly enough and also whether enough people can afford them. Considering that climate change is a catastrophe that is technologically induced and financially driven, we have to answer another key question: is another financially driven technology the way to fix it? What about just using less energy in the first place, cutting back on human activities such as travel, growing our own food, or getting serious about reducing both population and consumption?

Given the measurable changes in the Earth's atmosphere and climate, an increasing number of scientists and politicians are on the same page at this point, with both groups calling for rapid, global reductions of carbon emissions—for a transformation, in fact, from societies based on a fossil fuels to ones based on renewable energy resources. This assessment of what to do first is very positive. However, we don't want to make huge commitments only to end up with cures that are arguably worse than the disease, like agricultural biofuels or nuclear power. Now that we know human activities can affect the progression of atmospheric change, anything we do will have to be very well thought out. Solutions will have to prove they slow down the process without leaving new forms of pollution, disease, or imbalance in their wake.

It helps to begin by understanding the scope of the problems humanity is facing. Although the problem is generally referred to as "global warming" or "climate change," what's really happening is a rapid increase in the proportion of several key gases, including carbon dioxide, in Earth's atmosphere. Only relatively recently have scientists been able to amass solid information about the atmospheres that envelop other planets in our solar system. Like the porridge in the story of the Three Bears, one planet is too cold, one is too hot, and only Earth is just right. Venus, wrapped in a thick atmosphere rich in carbon dioxide and sulfuric acid, has a surface temperature of 450 degrees Celsius (843 degrees Fahrenheit), which would melt lead. Mars's atmosphere is too thin to trap heat and fluctuates from freezing cold (–120 degrees Celsius/–180 degrees Fahrenheit) at night to boiling in the daytime.

All life-forms on our planet can be said to have been created from what the ancients called the four elements: air, earth, water, and fire. We cannot exist without every one of them, and so far we haven't found any other place besides Earth that provides all four. When humans learned to control fire, it allowed us to get at the energy in our food more efficiently, frighten away major predators, and move into colder areas of the planet. But discovering how to control fire was also our first step in releasing the carbon that had been stored in peat, wood, animal oils and dung, coal, and other fuels, putting that carbon back into the atmosphere. The web of photosynthetic organisms on land and in the oceans can adjust over time to absorb excess or sudden pulses of carbon dioxide. So, despite human use of fire and the occasional meteor strike or volcano, the evolution of the atmosphere continued naturally through varying levels of greenhouse gases until the First Industrial Revolution, a mere three hundred years ago.

Then people began to make machines that consumed fossil fuels, and that mechanization started to reverse the entire process begun by plants so many millions of years earlier. Instead of sequestering carbon dioxide in living things or burying it safely below the Earth's crust, our machines do exactly the opposite. Bringing carbon-based fuels like coal and oil up from under the Earth and burning them releases the formerly sequestered carbon dioxide into the atmosphere. Unfortunately we hadn't recognized the systemic difference between this type of energy and the energy we had derived from natural sources, like the sun, animals, and running water, when we enthusiastically made this atmosphere-destroying process the basis for modern industrial society.

Today human activity is generating more than twice as much carbon dioxide as the biosphere can absorb, thereby elevating its concentration in the atmosphere above preindustrial levels by more than a third,[2] as well as acidifying all the oceans (carbon dioxide combines with water to form carbonic acid). Despite recent attempts to control CO_2 emissions, they're continuing to rise at the rate of more than 5 percent per decade and are responsible for about 60 percent of global warming. According to climatologist Andrew Weaver, bubbles of air trapped in successive annual layers of Antarctic ice reveal that "Our level of atmospheric carbon dioxide is now substantially higher than at any time"[3] during the previous 800,000 years. All fossil fuels raise carbon dioxide levels, but some are worse than others. Coal generates the most carbon dioxide per unit of energy, making it the "dirtiest" fuel. Petroleum pollutes the atmosphere 20 percent less than coal, and natural gas is the "cleanest" fossil fuel, putting out 40 percent fewer pollutants than coal.[4]

Carbon dioxide is not the only gas we need to be concerned about. Methane is an even more potent greenhouse gas, and we generate lots of it in our huge garbage dumps and landfill sites, in rice paddies, and in the flatulence of our domesticated ruminants, like sheep and dairy cattle. Vast, previously sequestered deposits of methane in now-melting permafrost in the Arctic or in rotting plants flooded by big dams are adding to the proportion of that gas in the air as the present global warming trend continues. Nitrogen oxides emitted by cars and used as fertilizers in industrial farming have also become a very significant source of greenhouse gases. Entirely new groups of greenhouse gases, chlorofluorocarbons (CFCs) and other halocarbons, created by the chemical industry for use in refrigeration and aerosol sprays, have a warming potential that can be as much as 24,000 times as powerful as carbon dioxide per unit of volume.

Currently transportation alone accounts for some 40 percent of all our energy use; that's partly because the number of cars is growing faster than the human population—worldwide, more than *ten times* faster. And instead of saving our trees to absorb all that carbon dioxide, we're cutting them down to make into disposable packaging that we use briefly then toss into carbon-exhaling landfills and incinerators. Every gallon of gasoline, when burned, creates more than 20 pounds of CO_2, an amount that would take a large tree about a year to absorb. It becomes mind numbing to compute the amount of gas we burn in our cars and the number of trees it would take to absorb that waste. When we consider the role of carbon dioxide and other gases in global warning, the crux of the climate change crisis becomes inescapably clear: we have been generating more greenhouse gases by burning fossil fuels, while diminishing our ability to remove them from the atmosphere, as we cut down trees and destroy other carbon sinks that could absorb that carbon dioxide.

Obviously we don't want so much carbon dioxide in our atmosphere that the planet starts resembling Venus. Evidence from recent probes suggests that Mars once had an atmosphere conceivably able to support life, so losing a functional atmosphere isn't an unthinkable danger, even in our own tiny solar system. We are in a predicament. On the one hand, there is our balanced atmosphere, without which most life cannot survive more than a few minutes. On the other, there is the entire fossil-fueled economic system humans have built over the past three centuries. If we are going to survive, the former has to outweigh the latter, and we are going to need to replace fossil fuels.

Tides, geothermal, wind, solar—each of these new energy sources has its drawbacks, but this fact should force a true knowledge revolution: we should never again trust single answers or single resources. It should give us great hope that no single energy source can be "the answer" that fossil fuels seemed to be in the First Industrial Revolution, since a diversity of small solutions is one of the key signals that a system may be sustainable. As well, energy sources more dependent on natural cycles, such as tides, force people to act locally and to manage that way as well. Iceland, with its cold climate and lack of arable land, would be crazy to ignore its wealth of geothermal energy and try to grow biofuels; North Dakota might possibly produce second-generation biofuels, but it would be better off looking into wind. Wave and tide power might be one way Indonesia could relieve the pressure on its forests, while solar is the obvious answer for dry, sunny places like Egypt or central China. Rather than rely on the fossil fuel monoculture that has led to wars, political destabilization, and massive corruption, we will have to develop a new polyculture of energy sources. The energy from most of these, thankfully, will be too little in quantity or too difficult to transmit to interest a big player. That alone should eliminate some of the scale problems any new technology might carry and should also help control the economic and environmental destabilization we're now experiencing.

We now have the technology to balance human needs with those of natural systems and are beginning to learn the methodology. Whether or not we achieve this aim obviously has a great deal to do with social values, economic anxieties, and political will. Although national legislation and global treaties with enforceable targets for carbon emissions are vital to this survival effort, the players coming up with the best solutions are not politicians, huge corporations, and economists. A remarkable number of solutions are coming from small municipalities, independent entrepreneurs, social collectives, and aboriginal or traditional societies. As with everything else sustainable, many answers as to how people will have the energy sources to help large populations survive economically—and retain a planet with a functional atmosphere—are coming from the bottom up.

CITY MICE AND COUNTRY MICE

What we are doing with energy is adding another cash crop.

Dan Juhl, community wind farm pioneer[5]

If we think of wind energy leaders, we think of Germany, Denmark, and maybe California, none of them known for howling winds. However, these places do have enlightened governments and citizens actively looking for ways to cut fossil fuel use. Today more obvious candidates in the wind energy sweepstakes are taking advantage of what such friendly government policies have pioneered, and acres of giant windmills are marching across the horizons of upstate New York and great plains states like Texas, Iowa, and Minnesota, not to mention Mongolia and Alaska, where ceaseless winds blow.

Right now in the new energy production center of southwestern Minnesota, rows of huge propellers hum with the power of an atmospheric energy resource that up until recently was viewed as more of a misery than a help. Farmer Marty Espenson says, "I hated the wind. It blows your hat off; it blows the seed corn bags across the field... you can't spray, it's too windy... But I like the wind now." Espenson and his wife, Patty, are among the first local farmers in Minnesota to profit from this renewable resource. They're co-owners, with eleven others, of the Bingham Lake Wind Farm, and they made their rather daunting initial borrowed investment back *in a single year,* by taking control of what is essentially a free resource. They've used the returns to stay on their land.

Dan Juhl, the wind entrepreneur whose nearby sixteen-year-old Woodstock Wind Farm inspired the Espensons, helped set up their cooperative and now travels the state helping other struggling farmers set up more wind co-ops. Along with his original 10-megawatt-producing farm, Juhl has assisted locals with the complex work of getting suitable land and then dealing with banks and equity agents to leverage loans. He's helped create 130 megawatts of community wind farms over the last few years. Some farmers only want a couple of these machines so that, like their German counterparts, they can supplement and diversify their smallholdings with the new cash crop of wind energy. Others have suitable acreage that they can combine with neighbors to form larger farms.

Wind power is like any other technology. It can have very good or very bad effects on communities and their natural systems. When wind farms are small, locally owned ventures that respond to the needs of the long-term residents, they tend to be good; after all, these machines consume no fuel to operate, and the electricity they produce is entirely emission free. Fossil fuels and other materials are consumed and emitted in the process of manufacturing windmills, but the structures more or less pay for themselves within a year or so through their long-term clean habits.[6] Still, we have to remember to weigh every installation and use carefully. Using windmills that rotate more slowly is mitigating damage to migrating birds, while moving the turbines to carefully chosen sites is helping to protect migrating bats, a species in alarming decline. And even though the global power potential of this resource is projected to be much greater than the entire energy needs of the planet, it's not going to be feasible or desirable to cover *all* our rural, offshore, and wilderness areas, however delightfully windy, with humming, rotating blades.[7] There are undoubtedly other natural system downsides we haven't noticed yet to go along with objections to their appearance and the noise windmills make. People living near them (presumably other living things are affected as well) report nausea, headaches, depression, anxiety, and sleeplessness so often that the complaint has been named "wind turbine syndrome."

These concerns aside, wind power could go a very long way toward solving some of this planet's biggest and scariest atmospheric and weather problems, especially if wind farms are funded through equity banks, government subsidies, and tax breaks that favor co-ops and community-based owners. We see this in the way Juhl helped his Minnesota neighbors navigate the intricacies of funding machines that cost $2 million apiece, setting local families up with the big investors who are in a position to apply for the government's complex tax credit schemes, so the providers of the land base this energy source needs can get in on its economic benefits.

Minnesota now leads the U.S. and Canada with its Hermann Scheer model of "indigenous power," and in a clear illustration that the groups really on the margins of energy supplies will embrace the best solutions first, India has shot up to being the fourth-largest producer of wind power in the world, producing 3 percent of the subcontinent's electricity from its free air. India's largest turbine manufacturer, Suzlon, has opened its first U.S. factory in Pipestone, Minnesota, near the Espensons' wind farm. Today the manufacture of rotor blades for windmills is providing three hundred long-term, highly paid, skilled jobs, and in any community, let alone a rural one, that's a whole new lease on the future.

India is not the only emerging country becoming a leader in energy entrepreneurship, however. China's wind power is growing faster than any other country's. Chinese leaders doubled their wind power goals for 2010 after reaching earlier goals three years ahead of schedule. Meanwhile, in the U.S., once a leader in a technology that became moribund during the George W. Bush years, people are getting back on the wind power bandwagon. The fact that wind power generation in the U.S. increased by almost 40 percent between 2006 to 2007 shows the possibilities are huge, and at the rate it's developing now in China, that country could bypass the U.S. in total production in the next few years. Given its alarming new consumption of cars, this is the most hopeful news of all.[8]

WAVES, TIDES, AND MOLTEN CORES

Ocean power functions in basically the same way as hydro-power... but rather than coming from river flow, this water power comes from the movement of the currents, tides and waves.

Global Warming for Dummies[9]

Geothermal energy... [is] a very large non-carbon-based energy resource and has the potential to be a significant contributor to the energy needs of this country.

M. Nafi Toksöz, professor of geophysics at MIT[10]

Tides and waves, unlike wind, are inexorable and extremely pre-dictable. All this energy and movement is something engineers understand in principle how to harness, using upside-down wind-mills or gates and turbines, and they even know how to store it. So why isn't there more ocean energy available? It's mostly a capital investment problem; the technology has lacked sufficient subsidies and government help. It's also what's called a "tenure" problem. It's not always clear who owns the oceans, especially if a fishery or other seaside industry is already in place. In many areas just getting the permits to start placing buoys and turbines can take years. The mechanical problems are challenging, too. Nasty and damaging as wind can be, oceans are worse. Storms destroy structures; saltwater corrodes; dragging cables and shifting anchors can tear up benthic fauna, the base of the food chain; and the marine power installations can disrupt migratory patterns of species like whales, birds, fish, and seals.

Although most of the ideas for marine power are still at the testing or construction stage, one of the world's first commercial wave farms did open off Portugal in 2008. Cornwall is developing a "Wave Hub" with a variety of wave energy converters floating offshore to connect up to 40 megawatts to the grid.[11] Another operating wave farm in the world is off the coast of Australia and uses "CETO" technology, which is a simple piston pump attached to the seafloor with a float tethered to it. Waves cause the float to rise and fall, and the pressurized water that results is piped to an onshore facility to drive generators. Not only is this the most elegantly simple design so far, it may (or, of course, may not) eventually prove to be one of the best. It's less likely to be damaged by storms and will not mar a beautiful seashore or attract birds or seals to their doom. However, if it ever became too popular we would have to think about what great numbers of pistons and cables would do to seafloor, where the entire food chain begins.[12]

Another energy source with huge potential is geothermal—and it comes from our interior sun, the molten core of Planet Earth. The magma flowing under our feet is 1,000 degrees Celsius (1,800 degrees Fahrenheit), more than twice as hot as the surface of Venus. It's more accessible in some places than others. The Philippines generates 17 percent of their electricity using geothermal energy and there's a lot of potential for the U.S., Latin America, Indonesia, and East Africa as well. New Zealand has thirteen plants and at least a hundred more possible energy plant locations, but although a few are locally owned by the Maori, others have been overdeveloped already or have high noxious gas content and other drawbacks.

So there are limitations, even in a planetary hotspot like Iceland, where the crust overlying the magma is thin and you can pump heat up very easily.[13] As Robert Zierenberg of the University of California at Davis notes, "Icelanders have taken as much as they can from their geothermal resource." Such areas do not produce indefinitely, either. The biggest geothermal development in the world, The Geysers in northern California, which provides over half of San Francisco's power, has begun to decline because "it was overproduced." It literally ran out of steam. Because of that experience, and for fear of depleting these unusual places where people have always marveled at the bubbling mud pots and steamy hot springs, many environmentalists have fought geothermal projects proposed for Hawaii and near Yellowstone.[14]

Geothermal heating can also be based on heat pumps that take advantage of the difference between temperatures above and below ground, a gradient that is the same the world over. Heating systems now common in Canada work like a refrigerator in reverse, heating buildings by pumping antifreeze-type liquids through a closed loop under the ground. This is another technology with a heavy initial capital investment that does eventually pay the investor back. However, with the present lack of good incentives in North America, payback can take fifteen or twenty years. Estimates on exactly how much energy all geothermal sources could provide range wildly, from the International Energy Agency's estimate of only 85 gigawatts—that is, not even 1 percent of humanity's current needs—to a 2006 MIT report claiming that the world's total resources, with technological improvements, could generate 100 percent of what we're using now for several millennia to come.[15] Violently diverging statistics like that are a feature of technologies still in development, because they don't yet have many big operational sites that can be assessed. So although the potential for geothermal power might be huge, we must remember that everything on this planet is finite, and the use of this energy is limited by many factors. Vents that once conducted dry steam can collapse, and wet steam reservoirs can run out of water. "Deep well" technology would bypass these problems by drilling hundreds of feet into the Earth's crust to "mine heat" from superheated magma. But like deep wells driven into aquifers, this idea sounds as though it could contain surprises. There are already observations that current installations can cause earthquakes, albeit small ones. Given Newton's third law of motion, "For every action there is an equal and opposite reaction," or the first law of thermodynamics, the fact that "energy cannot be created but can only change forms," drilling under the Earth's crust to extract heat could quite possibly be harmful to that crust or to us.[16]

Kevin Rafferty, formerly of the Geo-Heat Center at the Oregon Institute of Technology, predicts that "geothermal is unlikely to be the single answer to our energy security any more than wind, solar, conservation, or fossil fuels will be. It should be part of a mix of strategies."[17] And like every technology in this mix, geothermal will have to be carefully regulated. Most current geothermal sites are in fragile wilderness areas. The roads needed to access geo-thermal energy can disrupt habitat, and the steam that pours out of geothermal plants can be "heavily laced with high levels of salts and sulfur compounds... toxic to aquatic wildlife."[18] If that sounds minor, remember it's the wildlife on this planet, down to the smallest biota, that creates the atmosphere in the first place. And although geothermal energy is emission free and enormously better than coal, oil, and gas, we can't afford to put all our precious eggs in just one basket. To put that another way, if it's too big to fail, it's too big to live.

THE FATHER OF ALL

O Aten, ordainer of life... thou hast assigned to every one his place, providing the daily food... thou hast made millions of forms from thy Oneness—cities, towns, villages, roads and river... O Living Aten.

Hymn to Aten, God of the Sun, c. 1370 BC[19]

If we did want to rely on one basket of energy, it would be filled by the sun. Using the sun for energy has a history dating back millen-nia. What we now call passive solar buildings were developed by Pueblo villagers in the ancient American southwest, by the Inca and the Maya, the Greeks and the Romans, and by kingdoms all around West Africa. This technique was refined by Hindu and then Muslim architects, who used the minarets of mosques as solar chimneys. A log house built only two centuries ago in Eastern Canada would have been oriented to catch every single ray of sun in winter, with western sheds blocking winter winds and upper windows welcom-ing summer breezes. Only when fossil fuels became so cheap did builders begin to blithely ignore ways to maximize the warming and cooling qualities of the sun's two major products: heat and wind. Builders today are finding that they must quickly relearn how to harness these two resources.

Whether in an active form—shining on a photovoltaic cell or a parabolic trough that converts heat to energy, or passively warming a building through large windowpanes—energy from the sun is, without a doubt, the renewable source that is the most promising, most basic, and least polluting of all. Really big solar installations covering many acres, like Planta Solar 10, a huge tower near Seville, Spain, could have deleterious effects on desert habitat if not carefully sited and monitored. But the most common solar technologies are usually no more than a few black panels on existing roofs or small industrial installations that can be placed in unused corners, even where old, dirty factories used to be.

Despite the technical, subsidy, and funding issues they still face, small solar installations don't mar the skyline, kill bats, make noise, or interfere with fisheries. Solar panel manufacture creates some serious chemical pollution problems, and that certainly has to be addressed. But beyond that thorn, solar energy is close to perfect. Yet of all the green technologies, it remains among the least developed, because governments still do not provide enough subsidies to offset the high capital costs of manufacture and installation; in fact, less than 2 percent of the subsidies the U.S. doles out to fossil fuel producers goes to all renewable energy sources put together! So solar—the ultimate power source that created fossil fuels in the first place—cannot compete, so long as every gallon of gas or oil that we burn is largely paid for by tax dollars.

In Germany the Geosol solar plant in the eastern part of the country is a good example of how easily solar can shine with some intelligent tax incentives, even in Germany's drizzly, gloomy climate. Geosol's 33,500 solar panels produce no noise. They're so easy to operate that the plant's only employees are one manager and two guard dogs. The plant sits on 37 acres of what was a major dumping ground for East Germany's notoriously polluting coal industry. Like so many of the poisonous wastelands fossil fuel use has created, the site has been unusable for agriculture or housing; now it's producing squeaky clean power. Even on a rainy day Geosol operates at 25 to 50 percent capacity, proving that such power sources do not have to be built only in Nevada or Mexico. The Germans have managed to build this capital-intensive plant and six others as part of their goal of 3 percent solar power by 2020. How did they find the money? Once again, it's Scheer's Law. As the *Washington Post* puts it, "The old-line utility companies [have had] to subsidize the solar upstarts by buying their electricity at marked-up rates that make it easy for the newcomers to turn a profit."[20]

Solar power use is slowly increasing all over the world. Besides big plants that can provide up to ten thousand homes with power, there are hundreds of more modest collector stations that just supply lighting or hot water to individual schools or apartment complexes in Japan, China, India, the U.S., Canada, and even West Africa and Mongolia. Solar has been absolutely booming in China for the past few years; 1,100 solar panels were installed on the stadium used for the 2008 Summer Olympics, and solar-powered streetlights now illuminate many Beijing suburbs. The 2008 economic downturn affected all alternative energy industries, because manufacturers of wind, tidal, and solar equipment weren't able to get their hands on bank loans to expand operations or finish projects. This reaction could not only postpone the vital move toward solar and other alternative energy sources; it could also force small, independent, and local companies into the hands of big utilities, who have a better chance of getting money from the banks. And that, as we've learned through hundreds of examples of large, distant owner mismanagement, is definitely not the way to go.

In China, however, even the economic downturn is unlikely to cool the domestic market for solar power. China has 30 million people—the entire population of Canada—who still live without any electricity in rural areas, where the mountainous terrain makes stringing wires impractical. Now that they can no longer depend so heavily on exports of manufactured goods to the rest of the world, China will have to use solar (and wind) power to grow as a country, even more so now that they're beginning to see the downsides of hydropower. Predictions are that 180 megawatts of solar power—which is huge—will have to be set up in China's rural areas. That alone will probably make China the solar power leader of the world.

There are many exciting new uses for the cheapest energy source of all—like canopied parking lots that soak up the sun through solar panels and then feed this power to the electric or hybrid cars parked below. The reasons such undoubtedly workable ideas remain mostly in the development stage are, as usual, economic and political. As long as oil, natural gas, and even ethanol are heavily subsidized and therefore cheaper, the far cleaner and more promising technologies of the old god Aten will not be taken up by the great powers. Fortunately, in the absence of a Scheer's Law for the whole world, millions of small players are stepping up to the plate and finding real climate change solutions.

SMALL IS BEAUTIFUL

Solutions are not coming from Washington—they are coming from mayors and their cities.

Manny Diaz, Mayor of Miami 2001–2009[21]

The basic problem of how to fund alternative energy projects was summarized by Amory Lovins, one of the world's most respected energy innovators, in a 2008 interview. "We have obsolete rules that favor big over small, supply over efficiency, and incumbents over new market entrants... De-subsidizing the whole energy sector would be a wonderful advance to level the playing field but also to let renewables in. The barriers that renewables and efficiency measures face come less from our living in a capitalist market economy and more from not taking market economics seriously, not following our own principles [of fair competition]." When it comes to letting in the really efficient players—small, local wind, sun, or geothermal co-ops and companies—Lovins has become cynical. "I think the important policies need to happen at a state rather than a federal level. With modest exceptions, our federal energy policy is really a large trough arranged by the hogs for their convenience."[22]

When national energy policies became moribund in North America during the Bush/Harper decade, an unexpected group suddenly emerged into a position of leadership. States like California, and provinces like British Columbia and Quebec, illustrated Lovins's point and started to take matters into their own hands. They quickly began to outperform federal agencies in terms of serious carbon reduction policies. At the same time, a groundswell of even smaller players—municipalities—ignored both state and federal foot-dragging and started to make waves (as well as windmills and solar panels) for their regions. The Mayors Climate Protection Center, founded by the United States Conference of Mayors in 2005, has surpassed even state and provincial initiatives and is now leading change on the administrative level across North America.

Seattle's mayor Greg Nickels, the current president of the Conference of Mayors, cut his municipality's greenhouse gas emissions to 40 percent of 1990 levels and says they're just getting started. Chicago has vowed to become the greenest city in the country in this century, a tall order for the old Midwestern nexus of the steel and oil refining industries. This is a city that takes warming seriously; hundreds of people died in a heat wave in 1995, and Chicago's average temperature has risen 2.6 degrees since 1980. Its program aims for 80 percent reductions from 1990 levels, has already built more green roofs than any city on the continent, and promises a 90 percent reduction of solid wastes, more solar and wind use, and a major expansion of the transit system. It's also bringing energy codes for buildings "up to international levels." Rebecca Stanfield, a senior energy advocate at the Natural Resources Defense Council, reported that Chicago's work is not greenwashing but is "very specific" and "broken down into achievable sections."[23]

The spirited group of mayors bent on such innovations started off with just 141 members but within two years had swelled to well over a thousand. U.S. House Speaker Nancy Pelosi met with them in 2008 to pick their brains about green jobs, which the coalition estimates already keep over 700,000 people employed and could be greatly expanded. Their mandate is not only to get their own houses in order, with carbon emission reductions, efficiency measures, and more green space, but also to lobby their state and federal counterparts to "meet or beat the greenhouse gas emission reduction target suggested for the U.S. in the Kyoto Protocol."

The Conference of Mayors is ahead of the curve on being suspicious of quick fixes as well. In June 2008 they adopted a resolution to *ban* Alberta tar sands heavy oil from their purchases and also to be wary of agricultural biofuels. That resolution calls for "the creation of guidelines and purchasing standards to help mayors understand the greenhouse gas emissions of the fuels they purchase *through their entire lifecycle,* from production through consumption." Mayor Kitty Piercy of Eugene, Oregon, said of Alberta's heavy oil, "We don't want to spend taxpayer dollars on fuels that make global warming worse." Thousands of cities (including some of the biggest on the continent) represent a very big market to have opted out of Canada's tar sands products. The resolution will also help inhibit development of other "dirty oil" sites in Russia, Venezuela, and the U.S. itself.[24]

There is now a Mayors' Hemispheric Forum as well, where David Miller, mayor of Toronto, was invited to speak; that's because he's the chair of an even bigger climate change coalition of the world's major urban centers, the c40 Cities Climate Leadership Group. From Athens, Houston, and Berlin to Beijing, Bangkok, and Addis Ababa; from Cairo, Delhi and Hong Kong to Buenos Aires, Caracas, and London, city governments are making it clear that a whole new way of talking about environmental action has opened up. These unlikely political allies are becoming the wave of the future, possibly because municipalities are the expression of democratic government at its most basic level. At any rate, they're the ones bypassing the usual diplomatic protocols, where only nation-states are allowed to talk to one another, and where long-standing grievances keep leaders from doing even that. These municipal groups are the ones coming up with standards, goals, and increasingly binding agreements in the current crisis. As they say on their website, "The battle to prevent catastrophic climate change will be won or lost in our cities."[25]

It's very clear to c40 members that to shrink humanity's carbon footprint, all cities must strive to be more like the European ones tourists so enjoy. Urban dwellers must be brought closer together, in denser groups that need fewer resources, especially in terms of transportation. They'll need to eat local food raised without chemical inputs; that means big greenbelts to protect their water, air, and local food sources. The citizens should be given good reasons to get outside and walk and get to know one another so that population density is enjoyable and there are alternatives to energy-based pastimes.

Every year the U.S. Conference of Mayors gives members awards for "City Livability." In 2008 Louisville, Kentucky, was a winner for its mayor's "Healthy Hometown Program," a long-term effort to "encourage and support physical activity, healthy eating, and lifestyles," which included hosting popular "worksite wellness conferences." These are humble but sustained efforts to help people enjoy life where they are, in the here and now—so that they don't have to jump into a car or a plane or plug into an energy-fueled fantasy world to feel good. The municipal attack on climate change has so many built-in sustainable qualities it almost makes one glad it took the feds and the big movers so long to get on board. Right now the locals and the little guys are way out in front in terms of organization and creative solutions. And history has shown they're the ones closest to reality, who also have the most staying power.

TAX BADS, NOT GOODS

When the ill effects of leaded gasoline became clear, Malaysia simply taxed it, creating an immediate, nation-wide shift to unleaded gas.

Editorial, *Ottawa Citizen*[26]

All levels of government use tax policies to reward an activity they're trying to encourage and to discourage less desirable activities. So citizens are taxed for smoking, fined for speeding, and given tax breaks for developing land or building industries. However, governments net most of their revenue by taxing personal incomes, personal property, company payrolls, and retail sales; and they also tax urban buildings on the basis of their size as well as their condition. Turns out that by doing so, they're levying fines against socially beneficial actions, such as employing more people and keeping buildings in good repair. If they were to do the opposite—that is, tax practices that are socially or ecologically *destructive*—not only would they send clear messages to polluters and greedy business interests, but they'd also be rewarding altered behavior.

The Organisation for Economic Co-operation and Development represents thirty of the world's richest nations, producing two-thirds of the world's goods and services. Even though this organization is one of the most outspoken proponents of the benefits of economic globalization, back in 2001 it strongly recommended a coordinated program to remove environmentally damaging subsidies and *introduce green taxes*, "to prevent irreversible damage to our environment over the next twenty years." The OECD's own computer simulations demonstrated that removing subsidies in OECD nations, imposing an energy tax linked to the carbon content of fuels, and taxing all chemicals could result in a 15 percent reduction in anticipated carbon dioxide emissions by 2020. The economic costs of this radical change? Almost negligible—less than 1 percent of the member countries' GDP in 2020.[27]

British Columbia's new carbon tax, at $15 a ton, is dwarfed by Sweden's, at $120, which is well above the Kyoto target. The usual excuse in Canada and the U.S. for not signing on to such agreements is that it will cost too much. But Sweden's economy is booming, having grown 44 percent since they introduced this tax, which goes to fund many other programs. Even when modest costs are incurred as people turn away from fossil fuels, they can be recouped simply by *gradually* removing all the incentives and subsidies currently enjoyed by the gas and oil industry and reapplying them as rewards to sustainable technologies. The citizen won't be paying higher taxes and businesses will benefit—albeit in a different and, as it happens, much more profitable way. What most people don't realize, especially considering the way we talk about minerals such as oil and gold, is that taking things out from under the Earth's crust not only poisons us and all living things—it's a losing economic proposition.

All minerals are finite and will eventually be depleted, even from the most ample deposits. The mining, chemical, and oil companies certainly know this, because they've begged for, and received, massive "depletion" tax breaks and subsidies to enable them to "stay in business." In the United States mining of even outlawed toxic minerals like asbestos receives a special income tax deduction called the "percentage depletion allowance." All investments in oil and gas enjoy a "passive loss" tax shelter in Canada as well. Solid waste incinerators release tons of toxics into the air, but their construction is often financed with bonds that are exempt from federal income tax.[28] These subsidies could be switched over to modern, energy-efficient technologies that do not currently enjoy any such advantages. Citizens everywhere have to call their governments on providing perverse subsidies that might become the death of us all—after all, when it comes down to it, we need an atmosphere on this planet a lot more than we need these outdated heavy industries. Fortunately there are some excellent policy alternatives already coming into play.

Cap and trade, a method to control carbon emissions and gradually fund the inevitable shift to renewables, is used in a variety of forms worldwide. Although common, it's a controversial method and involves equally controversial "emissions trading." A government sets an overall emission target or "cap" on certain industrial wastes, then companies are granted allowances or credits to emit these pollutants, after which they have to pay for the privilege. Payments go directly to the government, or are given to any industry that pollutes less than the cap. In theory, the buyer of credits is paying for polluting, while the seller is being rewarded for reducing their emissions below the cap.[29]

The Kyoto Protocol was essentially a cap-and-trade system, with countries agreeing to meet certain caps in a set amount of time or pay a forfeit. The key, of course, is the price of the credits and how they are allocated. When they were given to national utilities in Europe for free, electricity charges actually went up because of a complicated market reaction, and some companies made windfall profits, all without reducing any of their emissions! Credits need to be publicly auctioned so that there's less opportunity for price fixing and manipulation. The cap—the limit on how much carbon dioxide you are allowed to produce—also has to be reset every few years, always downward. Not only is this system good for the atmosphere, but it provides steady tax revenues; a typical rate for the U.S. would be $20 billion to $30 billion a year, which would help pay off all those bank bailouts. However, this money should first go toward an earned income tax credit to adjust for the admittedly regressive effect of the cap-and-trade method of taxing. Otherwise, it penalizes poorer people disproportionately, as more expensive fuel is still a necessity and takes a larger bite out of small incomes.

There are scads of refinements—offsets, for example, that require a lot of expensive regulatory oversight to be sure they're being applied correctly. They're basically a voluntary tax, say on an airfare. The concerned person pays a certain amount to an international broker or trader so that the carbon they figure out their trip is creating will be offset by a proportional monetary support of "green" technologies—say disbursed to wind, solar, or geothermal installations. They've also been awarded to tree-planting or greenhouse gas–destroying programs, although both of those have been very problematic. Without serious government oversight, a flood of unscrupulous companies jumped into this market, which either didn't do what they said they would or whose policies actually caused perverse reactions.[30] The details of cap and trade are too involved to discuss here. Organizations like the Climate Change Coalition or the David Suzuki Foundation can supply analyses of all the issues online or on request. One thing everyone agrees on, though, is that there can't be "free credits," as there were at first; and prices for credits have to be fairly high or industries won't be affected and won't make the new technology investments necessary to switch to renewable energy.[31]

When all is said and done, some environmental NGOs and concerned scientists distrust cap and trade almost as much as business does—and for similar reasons. The problems with cap and trade partially lie in its virtues: flexibility and vagueness. As *Toronto Star* writer Tyler Hamilton puts it, "it's complex, lacks transparency, is vulnerable to manipulation, and creates a new layer of bureaucracy— filled with experts, lawyers, traders, and accountants—that's likely to result in bloat and inefficiency." These qualities greatly increase the possibility of the entire system becoming corrupted. Hamilton adds, "It also doesn't do much to influence consumer behavior, because the financial penalty [the industries] pass on is hidden. Out of sight, out of mind."[32] Even one of its major proponents, the Pew Center on Global Climate Change, admits that "although a critical and effective component of any comprehensive solution to climate change, cap-and-trade programs alone cannot achieve the greenhouse gas emission reductions required to stabilize the climate."[33] It is, however, still an important, market-friendly mechanism to use along the way.

The real point is that governments need to levy taxes of some kind that make spewing carbon more expensive, whether that goal is achieved directly or through cap and trade. There are lots of other mechanisms that will help, like raising fuel-efficiency standards (seen as cheaper and more acceptable to consumers), as well as using those subsidies that used to go to fossil fuels to gradually build up alternatives. Business-friendly conservatives like Republican congressman Bob Inglis and Arthur Laffer, who used to sit on Ronald Reagan's Economic Policy Advisory Board, have gone so far as to say that a "simple carbon tax coupled with an equal, offsetting reduction in income or payroll taxes" could, as Tyler Hamilton paraphrases, "lay the foundation for a dynamic U.S. energy security policy both Republicans and Democrats could support." Which puts them firmly in bed with the likes of Al Gore.[34]

DOUBLE DIVIDENDS

Kerosene, you are burning, it's gone. Only remaining the smoke and ashes. But solar energy you get it every day, again and again. This is called renewable. It's unlimited resources.
Dipal Chandra Barua[35]

One way to tell if a production technique or management approach is healthy for living things and can be sustained is to see whether it produces double or even triple dividends, as opposed to benefits that have to balanced against "trade-offs." Burning oil or gas cooks your food, but it costs a lot and pollutes. Using solar energy is far cheaper, causes no environmental degradation, and still cooks your food. When this idea is applied to taxes, this picture emerges: for every percent that a factory is taxed for its carbon emissions, that factory is allowed to pay 1 percent *less* tax for its employees. The workers still get their full benefits because the government will still be collecting taxes. The factory will be motivated not to fire more people but rather to cut down on emissions. Such double-dividend taxes are already in force in Germany, Spain, and nine other European countries. Spain's gasoline tax yields an extra $10 billion, which the country uses to offset the loss from cutting social insurance tax rates on employers and employees. In Germany over 1 percent of social insurance payments are being offset by carbon taxes; it joins countries like Holland, Norway, Denmark, Belgium, Switzerland, and Sweden, which already have similar taxes in effect.

Other kinds of taxes can be converted to double-dividend benefits. For instance, taxes on gasoline, especially in North America, are very regressive in that they cost the poor proportionately more than the rich. Because public transport is virtually nonexistent in much of North America, many people have to get to work by car. If they make very little money, giving them payroll or income tax breaks doesn't help much. The way around that, says Kai Schlegelmilch, a specialist on green taxes who works for the German federal government, is to provide gas tax rebates to people under a certain income. Even better, they should be given serious rebates on more fuel-efficient vehicles. The money gained from the richer drivers should go toward helping the poorer ones during this transition period. It should also help establish long-term solutions like train, bus, and trolley services. North America could adopt the highly efficient jitney services that are found in the Third World—small trucks or buses that load up on passengers and ferry them from town to town—but with more comfortable vehicles. Ultimately, if towns, cities, and suburbs are designed so that people can walk or ride together to buy groceries and run their errands, they will. More intelligent urban planning would also provide the bonus dividend of getting the population's weight down and improving general health. Reinstating services like home delivery of groceries and milk, using online purchasing, would put only a few commercial trucks on the road instead of thousands of personal cars. Double-dividend taxes can work for the poor, even in places that are inhospitable to anything but car travel.

When it comes to phasing out fossil fuel use in transportation, the good news is that we have many options, largely because the automobile is probably the most massively subsidized technology in human history. United States taxpayers are subsidizing drivers to the tune of at least $300 billion a year! As authors James MacKenzie, Roger Dower, and Donald Chen point out, "Motorists today do not directly pay anything close to the full costs of their driving decisions. However steep the bills for cars, insurance, automobile maintenance, and gasoline may seem to drivers, federal and state policies spare them many other costs. The net effect of these policies is to make driving seem cheaper than it really is and to encourage the excessive use of automobiles and trucks."[36] We all pay the costs of traffic congestion, lost time, lower worker productivity, and increased maintenance. On average only a little over half of what we spend on roads every year is covered by gas taxes and user fees. The rest is paid by people who may not even have a car. Getting rid of these perverse subsidies gradually, as many states and municipalities are doing, while immediately investing those funds in better forms of public transport, means that phasing out car use could be economically painless.

Germany shows how a different set of subsidies work. Train travel is subsidized by taxpayers there. This support has enabled German light and heavy rail to be seamlessly connected at shared stations so that people can move from tram to bus to subway to intercity trains with extreme ease. All trains have racks for bicycles and large packages, even lockers to store things in the stations. A German citizen can get around a typical city on a rail pass costing about US$50 a month. It enables the holder to use four systems: the U-trains or classic subways; their connecting buses; the S-trains that run between cities; and in some cities, Berlin for one, charming electric trams as well.

Fast, efficient, reasonably priced train service like that is ubiquitous in Europe. Here in North America, we're busily tearing up our rail beds and giving away rights-of-way just at the moment when it's vital to be expanding every type of train. It's a disgrace that neither the United States nor Canada has anything remotely like Europe's superefficient, high-speed service; Spain has had trains like the Talgo, which outclasses anything we own, for half a century now. Germany is a very small country compared with Canada, of course, and can afford to provide rail transport to its entire territory; but Canada and the U.S. could, at the very least, set up rail service in their big megalopolis areas. The Windsor-Toronto-Ottawa-Montreal corridor and the Montreal-Boston-Halifax loop could so easily have high-speed, comfortable train service, saving the lion's share of personal and business car and air travel every year. So could the corridor stretching from San Francisco up to Seattle and Vancouver. Rearranging subsidies and new green taxes could push those annoying new beggars, transportation giants like GM and Ford, into investing in trains. Trains can provide mass transportation that offers everyone options. Most importantly, they can be adapted to be almost completely nonpolluting.

Small efforts like these have to mount up to make a real differ-ence, and there are some places where they can mount up fast. Most people have heard of the microcredit movement started by Muham-mad Yunus, who won the Nobel Prize for it in 2006. Thirty years earlier, this middle-class university professor, tired of the terrible poverty around him in his native Bangladesh and despairing of government and international efforts to alleviate it, made loans of $27 each to forty-two impoverished women so that they could start small businesses. Today he presides over the single most success-ful community finance bank in history, the Grameen Bank, which disburses $60 million to $70 million every month to a total of 7 mil-lion borrowers. This bank, unlike mainstream ones these days, is entirely self-sufficient. No government or private aid comes to Gra-meen, which employs 23,000 people. It continues to expand, with between forty and sixty new branches opening every month. No wonder. Unlike the big financial banks and mortgage lenders, Gra-meen enjoys a 99 percent repayment rate and doesn't have to pay its top executives millions.

That level of success doesn't seem to have been enough for Yunus, so in 1996 he founded Grameen Shakti (which means "force of power flowing through") and set off to convince destitute rural Bangla-deshis that they could afford clean, nonpolluting power in their homes. Grameen Shakti sells a 20-watt system that uses solar pan-els to power one compact-fluorescent lightbulb and three LED bulbs, interior light sources that are much safer than burning expensive kerosene—which emits smoke and puts people at risk of catastrophic fires in their thatched huts. The buyer makes a small down payment and buys the little panel system over the next three years. The hard-est part is convincing people they really can afford solar electricity. But Grameen understands that the way to talk to households is through the women, so it sends women vendors and technicians to sell and install its products. Yunus says, "Not even 1 percent of the borrowers in conventional banks are women; she is given to believe she is nobody." Of course, it is these nobodies who run each home and who see to all the necessities: food, clothes, schoolbooks for the children, and light to read them by. Three thousand new systems are being sold every month, hundreds of thousands of solar panels spreading across the country and into India.

Another venture, the Grameen Technology Center, trains women to provide technical installation and upkeep of the solar panels, along with methods to transform the ever-present cow dung of Bangladesh villages into biogas. One owner of a large poultry farm produces enough biogas from her thousand chickens to supply not only her own cooking needs but also those of four neighbors and a commercial tea stall. The tea vendor, instead of spending 3,000 taka (about US$50) a month for kerosene, is spending only 700 taka (US$12), while reducing his carbon footprint. Because Yunus has intimate knowledge of the knife's edge of poverty these people live on, he knows this is the best place to get huge returns on every dollar. He bubbles with high social as well as environmental hopes. By saving this money, which to us is such a small amount, he explains, "a family can improve their home, send another child to school," even save up to pay for a dowry without mortgaging their future to a money-lender. "You are not only working for yourself, but you are saving the environment, you are saving money, you are saving the future!"[37]

Ultimately, the way to judge the true economic as well as environmental viability of an industry is to imagine if it could be profitable if governments removed all subsidies. Could the largest number of people make a decent living over the long term in the transport industry if they were building and maintaining mass instead of individual transit? Quite probably. If we subtracted the real losses from clear-cutting the boreal forest and tearing up its soil for tar sands oil, could Canadians make as much building solar tower systems? Oh yes, and no doubt with lighter hearts.

Ironically, nearly all of the polluting industries that contribute most to climate change—oil, industrial farming, chemical, and transport—have been unprofitable for years and have had to be propped up by laws, subsidies, and wars. Green economic activities like small fisheries, organic farming, or community wind farms have managed to survive and grow—even if just barely—usually entirely unaided and often despite huge obstacles. Their ability even to survive clearly demonstrates that their practices are more sustainable. Most of the sustainable businesses we discuss in this book have been forced by the current industrial paradigm to operate on a very unequal playing field, where their direct competitors are massively subsidized and many laws intentionally work against them. Until double-dividend green taxes and other government help are more widespread, the only way new technologies like solar or wind power can really get going is if we level that playing field.

THE WICKED WEB OF REAL LIFE

The worried man in front of [Vetinari], who was so considerate of life that he carefully dusted around spiders, had once invented a device that fired lead pellets with tremendous speed and force. He thought it would be useful against dangerous animals. He'd designed a thing that could destroy whole mountains. He thought it would be useful in the mining industries... Vetinari shook his head ruefully. It often seemed to him that Leonard,

who had pushed intellect into hitherto undiscovered uplands,
had discovered there large and specialized pockets of stupidity.

Terry Pratchett's description of the high-tech inventor

"Leonard of Quirm"[38]

There are many fascinating, creative, and already functioning
technologies that address climate change by reducing carbon emissions. The trick is to recognize "solutions" that won't make anything
worse and will work over the long term, especially with the dizzying
choice of high-tech products being presented as humanity's salvation every day. One rule of thumb: very often the more unabashedly
techie a "solution" is, the less likely it will be ready for market and
the more likely it will remain part of that dreamy realm of cold
fusion and perpetual motion machines. People love to play with
new ideas and materials and to believe, as Paul Roberts puts it in
his "Seven Myths of Energy Independence," that "some geek in Silicon Valley will fix the problem."[39] However, natural systems are
bound up with countless variables and intertwined with unknown
numbers of other natural and infinitely complex systems. Although
many technical fixes we hear about could eventually be great, we're
on a stopwatch to address global warming immediately, and most
of these technologies are too vague and untested to warrant heavy
investment right now.

Carbon sequestration, for example, also called carbon capture
and storage, is what tar sands companies like Suncor claim will
take care of all the carbon dioxide that is emitted to get one gallon
of oil out of Alberta. It's also what the coal lobby says will make U.S.
coal burning "clean." But it remains a theory, unlikely to become
useful before 2030 at the very earliest. This is very late to be able to
address the worst impacts of climate change. Besides being expensive and probably doubling the plant costs for a coal-fired utility,

up to *40 percent* of the energy that utility produces will have to be used just to inject the carbon underground. And researchers have no idea whether industrially sequestered carbon will stay where humans put it. "Even very low leakage rates could undermine any climate mitigation efforts."[40]

To quickly deal with the nuclear industry, which has wrapped itself in the green cloak of "no carbon emissions" and is attempting to stage a comeback: it's an expensive, dangerous, and resource-hampered nonstarter. First of all, every initial stage of nuclear production produces huge amounts of carbon emissions, except for the final fission. So like big hydro, it's not really clean. Uranium enrichment requires halogens, greenhouse gases that can be up to ten thousand times more potent than the carbon dioxide they are supposed to reduce. Although nuclear power has been used to fuel France's relatively low-emission lifestyle, the price even for the French has been high: constant discharges into their own waters and the English Channel, mountains of radioactive waste badly stored with no plans for the future. But the real clincher against nuclear energy is the limits on future expansion simply because uranium ore has become a very rare and expensive commodity. Like fossil fuels, nuclear fuel is finite, and that fact, as we should have learned, creates serious problems if we base our economies on it! Renewables are sustainable in fundamental ways nuclear power can never match.[41]

A promising modern technology involves using wood lignin and other waste substances for second- and third-generation biofuels. These could in fact be very beneficial for the atmosphere, if obtained in specific ways from certain industries and localities and if carefully regulated. Chile, for example, is excited about the idea of forest waste–derived biofuels, but the definition of the word "waste" is key. Everybody is scrambling to get away from agricultural biofuels like ethanol, but the devil is in the details. Chile's plan mentions "farming and and forest waste" but also wants to get material "from *plantations* of poplar, paulownia, and acacia trees or maiden grass." Several leading forestry companies, two government consortiums, and two universities have been given $10 million to start a program and are developing protocols. This project concerns environmentalists like Daniela Escalona of the Action Network for Social and Environmental Justice. She fears the fast-growing tree and grass species the government favors would be cultivated at high density and would use up soil nutrients and water at a rapid rate. And since there's not much marginal land available in Chile, once again that space would likely come out of the little bit that's needed by poor or indigenous communities.[42]

A more promising biofuel technology involves growing algae in self-contained silos that get heat and nutrients from industry. The idea is to take polluted water full of nitrates—say runoff from an industrial farm—and use it to grow local algae species that thrive in polluted water. This is a very interesting form of biomimicry. The Redhawk Power Station in the Sonoran desert of Arizona is already equipped with greenhouses that grow algae nourished on its carbon dioxide wastes. Although very small, algae are "the fastest-growing plants on earth... the most adaptable, and... the richest in high-energy oils ideal for making biodiesel." They are also "the most efficient converters of carbon dioxide to oxygen and biomass."[43] In early experiments at MIT, Isaac Berzin, a chemical engineer and cofounder of GreenFuel Technologies, the company working with the Arizona Public Service at Redhawk, used algae to cut the university's cogeneration plant's carbon dioxide emissions by about 70 percent and its nitrogen oxide by 85 percent. However, we have yet to see any working systems that demonstrate if these systems would ultimately put less carbon dioxide in the atmosphere. Other companies are trying to genetically engineer algae to express the traits they're looking for, but Ray Hobbs of the Arizona Public Service says, "Given the dangers of unleashing a GMO (genetically modified organism) that adaptable and prolific... with twenty thousand [natural] species to choose from, why would you need to engineer a new one?"

Like the biotech approach of growing algae, many fancy, high-tech fuel solutions depend on sketchy developing technologies that we already know might end up being fully as dangerous to life on Earth as nuclear power or burning fossil fuels. For example, the wonderful, wafer-thin, paint-on materials that could make solar panels more common are largely created by using nanoparticles. Although nanotechnology shows great promise, it is still in its infancy.[44] The microscopic creations of nanotech could infect organisms and damage tissue and organs or interact with natural microorganisms in ways we do not yet understand. If history has taught us nothing else, it's that uncritical enthusiasm for new technologies without concomitant suspicion and careful study can get us into trouble.

When DDT was first developed, the researchers at Monsanto and Dow didn't know about biomagnification, the process that carries toxins up the food chain. When bombs were dropped on Japan, even Einstein hadn't imagined there could be radioactive fallout, electromagnetic pulses, or nuclear winter. The primary way to deal with a problem that is basic to scientific inquiry is to follow the "precautionary principle," an international guideline adopted by the UN that demands that new technologies prove they won't do harm, rather than leaving their victims to prove the reverse later on. All researchers need to become more proactive about possible adverse effects and operate under the behest of independent monitoring agencies that have the power to suppress new technologies when necessary. So far regulators have been leaving that policing up to the producers, whose profits depend on their product's continuance, regardless of possible dangers. Legislation needs to be put in place before a technology is commercialized. Once people start investing on a mass scale, it's much harder to back away.

Many of the advanced technologies for alternatives to fossil fuel use, like breaking down the cellulosic waste of potential biofuels, depend on genetically engineered (GE) enzymes, bacteria, and viruses. Although some regulation on genetically engineered organisms exists, legislation governing their creation and release is still in its infancy and is regularly breached. One day, one of the many reported contaminations of our food crops and food products—by experimentally grown genes that express medicines or pesticides, for example—will cause very permanent damage, if not to a few unlucky individuals, then more catastrophically, to entire plant or bacterial genomes. Already most of the historic landraces of Mexican corn have been contaminated with GE varieties, which means we can never go back to the source of this food plant to find disease resistance or other innate qualities. GE technology also continues to operate within a huge cloud of coercion and corruption. The industry is so politically influential that the U.S., in February 2009, was the first country in the world to announce it would *tolerate the contamination of food crops with unauthorized GMOs*,[45] some of which could be pharmaceutical and have unknown effects on human health. This is an admission of failure in terms of our efforts to anticipate and control technological dangers and a good illustration of how careful we have to be as we become increasingly technologically adept.

If the crisis shaking every natural system on the planet is to be brought under control, it will not be through solutions that help the atmosphere but threaten another fundamental ecosystem service.

Just because a technology is exciting doesn't mean it might not be dangerous. We are only beginning to reassess what kind of energy we should use in place of the fossils we dig out of the ground. Even if we manage to choose safe technologies that get proper investment, we can't predict which will be truly sustainable innovations. Batteries are undoubtedly going to be a critical area, and hydrogen fuel cells still offer potential, despite their failure to deliver so far. Solar power is largely good news. Wind and geothermal power are good news if locally managed and restricted. And biofuel from algae and tidal power are worth serious study, although we don't yet know how much or how quickly they can be scaled up enough to make a difference. It will take many years of research before we can find some of these things out. Meanwhile, the challenge is to reduce emissions *right now*. So what's the best answer to that?

THE REAL BANG FOR THE BUCK

Number of [U.S.] jobs created by spending $1 billion on:

Defense:	8,555
Health care:	10,779
Education:	17,687
Mass transit:	19,795

"The Page That Counts," *Yes! Magazine*[46]

The most sustainable, most predictable, safest, and easiest way to slow climate change is to cut the production of carbon dioxide in the first place. This is termed "energy efficiency," "energy conservation," or "nega-watts" and is supported by almost every climate change spokesperson, researcher, NGO, and activist in this field, from international government agencies like Intergovernmental Panel on Climate Change, the G8, and the OECD to public experts like Bill McKibben, Elizabeth May, and George Monbiot. Andrew Fanara of the U.S. EPA says, "it's the cheapest, cleanest, fastest way to act," and will be "the 'bridge fuel' we must build on and must invest in."[47]

As journalist Paul Roberts puts it, "Saving energy is almost always cheaper than making it,"[48] however clever or clean the new source. It's also much more economical to invest in the conservation process. From a social point of view, this shouldn't be so hard to do. A 2007 BBC World Service poll of over twenty thousand people from twenty-three countries found that huge majorities of those people, from a low of 60 percent in India to a high of almost 90 in China (with the U.S. at a respectable 78 percent), *favor taxes* to develop clean energy and to discourage the use of fossil fuels. Even higher numbers, especially in the First World, were "ready to make significant changes to the way they live to reduce climate impact." It looks as though we have only five or six years left to get programs online to address the most serious problem in human history. But that demonstration of mass willingness to get to work is the key to success, as well as a reminder of how far behind the curve most governments and international institutions and treaty negotiations are.

Just stopping simple leakage and waste would save nearly 50 percent and sometimes more of the energy we use to heat water in old tanks, warm badly insulated buildings, or power needlessly heavy cars and trucks. That's what we could gain even without introducing better, lighter cars and more insulated, sun-friendly buildings. But energy is not just being wasted by inefficiency and profligacy; it's being wasted on unnecessary toys and luxuries. We will have to learn to do without the chaotic, product-oriented life common in industrialized countries—being transported everywhere by engines and continuously plugged in to entertainment devices so that we need never experience a single moment of boredom. By 2020 electronic gadgets like iPods, cell phones, and computers will shoot past cars and trucks as the commodity using up the highest amount of finite energy resources.[49] They seem so self-contained and clean, but their components come out of the Earth and generally cannot be reabsorbed. They all ultimately depend on electrical energy, currently mostly generated by burning fossil fuels, to make their pictures and music. Some companies are taking initiatives to try to solve this problem; for example, Apple has recently released a recyclable MacBook shell that uses a quarter of the power to run and none of the usual mercury in the manufacture. But the current endless proliferation of new gadgets whose manufacturers accept no responsibility for their own waste stream—that is, who dump their environmental costs on society—is not going to be a viable way to survive.

Climate change confronts us with the need to see the whole picture—the interconnections and interdependence of all parts of the biosphere, from soil biota to ourselves. Shahid Naeem, a celebrated biodiversity researcher, writes, "Earth formed at just the right time, at just the right distance from the Sun, and with just the right kind of axial tilt to generate seasons, with just the right kind of moon." Earth was also bombarded with just the right kind of comets to deliver the elements to form water and the other necessities of life, along with the right kind of gravitational force to enable the thin skin of our atmosphere to insulate us like a blanket. Naeem goes on to say that the real purpose of human life on this planet is "stewardship... to ensure the proper biogeochemical functioning of the biosphere."[50]

Humanity used to know, and needs to learn again, how to live within the planet's inescapable geo- and biochemical boundaries. But we don't have to *feel* limited when we do so. Many of the changes in the way we live will involve valuing and enjoying what we already have, rather than yearning for what we don't. For example, preserving all the precious carbon sinks we have left, including peat bogs, prairies, forests, and wetlands, is an absolute survival necessity. This activity does not involve trying desperately to get something back—although some of that work lies ahead—but simply stopping practices of deforestation, burning and flooding, massive clearing to grow oil palm, soy, or beef monocultures, or just those little increments of paving over yet another local forest or wetland for urban or commercial development.

Every effort to preserve a wetland or a local habitat, buy pieces of Amazonian rain forest, or fight encroachments on parks leads to good news. There will be more good news when governments meet and then go beyond their existing commitments to put certain amounts of their territory (generally around 15 percent) under protection. It will multiply again when more of them pass legislation rewarding farmers who don't destroy forests, wetlands, or woodlots with tax breaks that will make saving a beaver pond or a wooded area economically viable. It's pure self-interest for rich countries to tie their international aid to support that will enable countries still rich in forests, coral reefs, or tundra, like Ecuador, Siberia, and Indonesia, to protect the entire world's supply of air and water. There is good news in the fact that as a species we're finally addressing the state of the planet's oceans and talking about how to protect what's left of its atmosphere- and climate-generating biota.

This stream of good news will continue for individuals engaged in this process. As the people who try to reduce, reuse, and recycle will tell you, living this way is simply better. It's more fun to contribute to life than to destroy it. Millions already know about sustainable solutions to the overriding problems we face and are encouraging others to make a difference: get a hybrid or a fuel-effi-cient car; take the bus or ride a bike; energetically support light rail and all kinds of trains; trade up to efficient appliances, especially your refrigerator; eat organic and local as much as possible; buy local and support all forms of businesses that cut down on size and transport; use those curly lightbulbs, turn down the thermostat, stay away from plastic bags and containers, recycle, reuse—and refrain from using in the first place.

The reason we've heard that "reduce, reuse, and recycle" refrain for years now is that such an approach provides the biggest, fast-est, and best bang for the buck. The biggest because consumers are signaling to markets that they won't support nasty products or processes and will support green products, as the vast increase in interest in public transport, organic food, and fair-trade lumber demonstrates. The fastest because this approach is as immediate as anyone's next meal or trip to the hardware store. And the best because people gain unexpected benefits, like getting fitter and making new friends that happen, when they, supposedly selflessly, give up eating so much meat, start walking to work, or join with their neighbors to save a wetland. Those already doing these things will tell you that their actions didn't turn out to be so selfless after all. Most of the truly effective ways to stop climate change are the simplest and most enjoyable. Their double dividend is the better society we build along the way.

LISTEN FOR THE JAGUAR

FORESTS

FREE LUNCH

Forest trees are getting bigger and absorbing more carbon.

Patrick Luganda[1]

Throughout the western United States, cherished and protected forests are dying twice as fast as they did twenty years ago because of climate change.

Stephen Leahy[2]

A forest is a lot more than just trees. It's an extremely complex community of living and dead organisms, minerals, and the elements of soil, air, and water; often it needs the fourth element, fire, to thrive. In fact, the fundamental building blocks of planetary life are completely interwoven in forest ecosystems. At the same time, forests produce and expand all of these components, so much so that for a long time they were considered to be almost the only carbon sinks—that is, the planet's prime producers of oxygen and consumers of carbon dioxide. Today we know that peat bogs, ocean biomes, and even prairies also provide this service. But when it comes to producing the greatest possible variety of life-forms and keeping the water cycle moving while simultaneously providing the planet with an atmosphere, there is nothing on Earth like a forest.

Most evolutionary theorists think humans began as tree-dwellers. Our ancestors then stood upright and walked out of the forests onto the African savanna to complete our evolution into *Homo sapiens*. We've never forgotten our leafy home base, though, and we keep going back into the woods to fetch most of what we need to survive: water from springs and streams, game animals, fruits, fungi, and nuts, building materials for our shelters, fuel to keep warm and cook our food, medicinal plants, and even feathers and dyes for dressing up. That list doesn't even begin to measure the real value of forests to us. For example, forests rich in tree species like hickory, maple, and beech don't just indicate rich soils, as the American and Canadian colonists thought when they claimed such stands as the best cropland. Those tree species are able to *create* rich soils because their thick canopies make for moist, well-decayed leaf wastes. The more open crowns of oaks produce thinner soils, while hemlock needles acidify soils.

But all trees unite in keeping the soils of the Earth from washing away in the rain. Forests absorb water and cleanse it; they protect and purify the water in lakes and streams. We've learned the hard way, only after forests have gone, that transpiration from tree leaves, as well as their ability to trap cloud and fog, creates a local area's rainfall. We have only recently learned that the presence of forests greatly softens climates—warming cold ones, cooling hot ones, mitigating wind and frost—to an extent never realized until we began to lose forests on the modern scale. All the elements necessary to the survival of the complex "higher" life-forms—soil, water, atmosphere, high-energy fuels, and a moderated climate—come out of the collective life that makes up our planet's forests.

Although we've always used forests as all-purpose shopping centers, one major benefit they provide that we also didn't appreciate until recently is their performance as carbon sinks. We've long known that they build and restore the planet's atmosphere by exhaling oxygen and that they capture carbon dioxide from the atmosphere as they photosynthesize. As long as trees are alive, they hold that carbon dioxide inside their cells while pumping out the oxygen that keeps our atmosphere in balance. But a recent megastudy funded by Britain's Royal Society and Natural Environment Research Council, published in the journal *Nature,* offers proof that intact, old-growth forests absorb far more carbon than we ever imagined.

An international team of scientists spent forty years studying old tropical forests across parts of Asia and Africa, from Liberia to Tanzania. They measured trees over those decades and were able to estimate stored carbon and average growth rates. They discovered that the trees in these forests alone are absorbing all the carbon humans are adding to the atmosphere every year by driving cars and trucks—that is, 18 percent of the carbon dioxide we create annually from burning fossil fuels. Moreover, these big tropical trees in intact forests are actually stepping up their rate of absorption. For the past few decades each acre of African forest has trapped an *extra* 0.24 tons of carbon every year. This increase may be due to carbon's fertilizing effect; at any rate, the trees are getting bigger at a faster rate as they "mop up" the plentiful carbon dioxide. Simon Lewis, the lead author of the study, says, "We are receiving a free subsidy from nature [that is] substantially buffering the rate of climate change," although at the rate this and other old forests are being felled or despoiled, that's not as much good news as we might think.[3]

What then seems puzzling is the equally recent news that although African trees are getting bigger on their carbon dioxide–enriched diet, North American trees are dying and never attaining their normal sizes and ages. Scientists think this effect may also be because of climate change. An article in *Science* that appeared in January 2009 quotes a study showing that in the undisturbed old-growth forests of the western U.S. and Canada there's been "a dramatic increase in tree mortality [that] applies to all kinds, sizes, ages, and locations of trees." The rate of death in northern coastal areas like Oregon and British Columbia has doubled in the past seventeen years, and in mostly frost-free conditions, as in California, it's doubled since 1990. The warmer temperatures brought about by climate change are being blamed; summer droughts are longer, the mountain snowpack is smaller and being used up earlier, and "warmer temperatures also favor insects like tree-damaging [pine] beetles." Tom Veblen of the University of Colorado at Boulder says, "We're seeing continental-scale evidence of warming. It is very likely tree mortality will increase further as temperatures continue to rise." So although climate change may have the temporary effect of helping the remaining tropical old-growth trees get fat and clean the air even more, an effect that benefits humans, neither trees nor people are getting the most calories out of increased carbon dioxide in temperate zones. Here, the free carbon lunch seems to be going to tree-munching pests like the mountain pine beetle and the spruce budworm.[4]

Why do trees thrive and grow bigger in the tropics and die in temperate zones? That puzzle is another of nature's complex mysteries that we are only beginning to notice and unravel, but it obviously has something to do with water availability and seasonal pest cycles. The Intergovernmental Panel on Climate Change reports that human activity releases 32 billion tons of carbon dioxide every year, but "only" 15 billion tons stay in the atmosphere as a heat-capturing blanket of gas. Of the other 17 billion, about half is being dissolved into our increasingly acidified oceans and the other half is being absorbed into land vegetation, bogs, soils, and woods. Of that, the researchers are now realizing, fully half is sucked up by the planet's few remaining tropical forests.

If we were to put a dollar amount on what that sink is worth on the carbon trade market, it would be in excess of $25 billion a year, a value that doesn't even consider the forests' roles as protectors of biodiversity and a key element of the planet's water cycle. Still, this latest discovery, that half of the globe's "land sink" for carbon is performed by the tropical trees, suggests a simple and elegant solution to all our problems. The Royal Society study's Simon Lewis says that rich polluting countries should be *transferring substantial resources to countries with tropical forests* to reduce deforestation rates and promote alternative development pathways."[5] This would be quite easy to do through existing wealth-transferring mechanisms like the World Bank. If the tropical nations stop cutting forests because we've forgiven the debts they're liquidating those forests to pay—we'll all breathe easier. Literally.

We're a long way politically from such a solution, but at least this new view of forests as a valuable asset when kept intact, as opposed to when cut down, marks a very positive development in recent human-forest relationships. It's still rare that economic profits are accorded to the caretakers of these incredible factories of air, water, soil, and biodiversity working around the clock, and by and large, the history of commercial timber companies, in terms of sustainability, is a pretty dismal one. We don't need to go into the centuries of slash-and-burn, cut-and-run, and clear-cutting that have rendered the planet's once magnificent blanket of green into a moth-eaten scarf. Liquidating the planet's atmospheric powerhouse of trees was done largely to get lumber to build things humans want, especially houses, and that's still going on. But one of the oldest companies in that business in North America has pointed the way toward a credible new future for all the rest.

WHAT DOWNTURN? WHAT PINE BEETLE?

Housing starts, our bread and butter, are down in 2009 to a quarter of what we had just three years ago. But we're gonna be the survivor, and that's all thanks to the sustainability part of our business.

Wade Mosby, Senior Vice-President, Collins Pine[6]

Collins Pine is a 150-year-old U.S. timber company that often turns up in discussions about what a sustainably managed commercial forest might look like. They operate what has been called the finest privately owned industrial forest in the country, and their practices have been praised by everyone from the Rainforest Action Network and the Sierra Club to the *Washington Post* and the *Christian Science Monitor*. Collins Pine employs 7,500 people directly, grossing about $250 million a year in plywood, hardwood and softwood lumber, oil, and gas. The business was started in 1855 by the present owner's great-grandfather in Pennsylvania, where it still holds 126,000 acres.

At the age of ninety, Maribeth Collins, widow of Truman, the man who set the company up to be sustainable, turned her chairmanship over to her daughter, Cherida Collins Smith. She's the key shareholder and oversees company practices. One son, Terry, lives in the remote village of Chester, California, where he manages the sawmill and oversees forest operations. Another son, Truman Jr., also serves on the board and is president of the company's busy charitable foundation. The day some years ago that we visited their big operation in Chester, on the gala occasion of cutting the two billionth board foot out of this thriving, diverse, sweet-scented forest, three of Maribeth's four children were in attendance. Maribeth Collins lives in Portland, Oregon, where the Collins family, staunch Methodists, has funded everything from libraries and scholarships to church construction and foreign aid programs. What's most remarkable about their company is their mission. They've formally pledged to do three things: maintain the health of the total forest ecosystem; support the production of wood on a sustained, renewable basis; and provide social and economic benefits to the surrounding areas and communities.

The proof that they mean it lies in two bottom lines: money and certification. Collins Pine makes smaller profits than its publicly traded competitors simply because its owners aren't greedy. In the 1990s it consolidated its traditionally sound practices by working with Paul Hawken, the founder of The Natural Step. Collins Pine decided to embrace TNS sustainability principles, becoming one of the first U.S. businesses, and so far the only timber company, to formally do so. That means every endeavor Collins Pine undertakes is weighed against short- and long-term economic impacts along with the four conditions of TNS: Does the action reduce the use of finite mineral resources? Does the action reduce the use of long-lived synthetic products or molecules? Does it preserve or increase natural diversity and the capacity of ecocycles? Does it reduce the consumption of energy and other resources?[7]

The company also voluntarily chose, early on and against some stiff internal protests, to become fully certified by the Scientific Certification Systems of the Forest Stewardship Council. Wade Mosby is senior vice-president at Collins Pine and a walking encyclopedia of company and local history. He says, "We're losing 25 percent of the kind of profit other outfits get with our more long-term method.

For example, the way we use natural tree regeneration means the forest matures at a more normal rate. The usual, even-age tree farm monoculture management style is 25 percent more lucrative over the short term. So we 'lose' hundreds of thousands of dollars that way. And for that reason alone, if we had a different family running things, we'd be long gone."

All across Canada, Russia, the United States, South and Central America, Africa, and Southeast Asia, most forest lands are either leased to or owned by huge timber companies. These companies still operate according to First Industrial Revolution logic. Because of the global market, most timber companies will insist that the only forest management model that will maintain their competitiveness and their profits requires industrial-style clear-cutting and the replanting of monocultured tree plantations. Governments bow to the wisdom of the marketplace and their own need for foreign exchange. The trouble is, few companies have existed for long enough to complete even one full crop rotation of mature trees; so they really are just guessing about how well their chemically sprayed, second-generation forest will produce over the long term.

Collins Pine has embraced a value system that goes beyond considering money alone, and, ironically, the company actually makes more by spinning out its harvest over a much longer period. This rational practice requires private ownership (no trading on the stock exchange) and self-restraint in the face of all those "market pressures." For example, when Collins Pine forester Barry Ford noticed that an area they were cutting in their Almanor Forest in northern California wasn't regenerating naturally and they weren't getting new seedlings, he realized it had been stabilized for so long—five hundred years or more—that the trees were no longer set up for reproduction. "That's when we started studying the microsystem. The rhyolite soil up there compacts, so we thought that was it; we tried to break it up and I've been analyzing all the nutrients. But until we figure out how to manage it, we've stopped cutting." It's still very useful to the company, Ford says. "Now it's become a study site to learn how this old forest managed to survive on such poor soils." In a comparable situation in Quebec—very poor regeneration of stabilized black spruce boreal forest between Lac St-Jean and Chibougamau—the lack of seedlings has only inspired the pulp companies to redouble their cutting.

The Collins Pine 94,000-acre forest at Almanor is principally sugar pine, magnificent trees that grow up to 180 feet high, with enormous, 2-foot-long cones that smell bewitchingly like hot syrup. This species suffers from blister rust, so Collins Pine only takes seedlings resistant to the rust for replanting. Ford says proudly, "Our sugar pine has carried this rust-resistant gene for a whole century. That's why you need a big gene pool. Nursery stock from China brought the rust into British Columbia in the 1930s in the first place, and it spread. We plant local seed only, and we make sure we keep all the genes that have evolved in this forest around, just in case." While other companies are jumping onto the bandwagon of the latest cloned supertrees and genetically engineered trees that express their own anti–spruce budworm pesticide, Wade Mosby says, "GMOs—oh my God! We're worried. Really worried about that. We'd never use those seedlings. We never even use clones. We use local seeds from the specific hillside or whatever we're planting on— only! Well, usually, we don't have to plant at all." His forester, Barry Ford, agrees. "We hate those fancy clones," he says. "They don't provide protection from new diseases. Fusarium just tore through the cloned Georgia pines that they spent all those years growing and raving about. More types of blister rust are just waiting out there."

Wade Mosby tells the story of a young man named Wally Reed hired by Truman Collins Sr. fifty years ago to help carry out a vision that had come to the young Collins. He had been trained as a forester at Harvard, where he first got the idea of turning his family's extensive landholdings into sustainable forests. Wade Mosby says no one was doing selective logging and local breeding the way Wally Reed was at the time, so he was constantly attacked by other companies, foresters, academics, and the industry. "He was completely alone, and he was only a young man in his early twenties! But he and Truman were so sure of what they were doing that they ignored all the experts, all the taunts and threats of bankruptcy. And Wally lived to see his work honored as the best commercial forest in the whole country, one that has *never* been degraded by bad practices." Wally Reed, frail at eighty-six but radiant with pride, was present at the festival marking the cutting the two billionth board foot at the Almanor Forest. Because Collins Pine listened to him, they still have as much wood in that forest as when he began managing it.

Collins Pine and other conscientious companies have benefited from the credibility they've gained through certification programs, a major source of good news. Today there are hundreds of new ways that consumers—from those buying coffee to those picking up two-by-fours—can tell whether a business or an organization really is a reliable steward of the natural resources it's exploiting. Particularly when it comes to forest products, the certification movement has become powerful. The Forest Stewardship Council (FSC), based in Oaxaca, Mexico, is one of the most well-known organizations overseeing sustainable forestry. They're a nonprofit umbrella that includes environmental groups, timber and trade organizations, foresters, indigenous people's organizations, community forestry groups, and other forest product certification organizations in twenty-five countries. They're pretty effective and reliable in northern countries like Europe, Canada, and the United States. But certification can also inadvertently create new markets in lawless areas like the Amazonian or Indonesian wilderness, encouraging a separate black market for the very trees we need to protect. So until the new organizations can deal with such problems in the Third World, organizations like Friends of the Earth are still supporting a boycott of all tropical wood.[8]

The whole system includes watchdog overseers, like Scientific Certification Systems, whose inspectors go into North American forests to make sure that their management is really measuring up to the FSC goals and that product integrity is being maintained all the way to the consumer. Near the Almanor Forest is southwestern Oregon's Rogue Institute for Ecology and Economy, which certifies small operations. It's affiliated with the SmartWood certification program. This program operates internationally to help private producers, especially small ones, get in on the expanding market in certified forest products. It's only been around for a little over a decade, but it has already led to fundamental and evolutionary changes, such as pledges from U.S. hardware giant Home Depot and Swedish furniture store IKEA to use only certified forest products in their stores. This alliance between certifiers, concerned consumers, and timber producers has translated into a serious market advantage for the few forestry companies with real ecological integrity.

The first Collins Pine forest to undergo a rigorous certification audit under Jim Quinn, president until 2000, was the Almanor Forest. In a proud and remarkably candid history of the company posted on their website, Wade Mosby is quoted as saying, "The foresters there were not happy. In fact... they were downright hostile. Bill Howe, who had just taken over as forest manager in Chester, went ballistic. It was an insult. It was an affront. And he was mad as hell." It was the early 1990s, however, and many environmental NGOs, from Greenpeace to Sierra Club to Friends of the Earth, had informed the public of the waste and rape of forests. People were demonstrating in front of construction supply outlets and consumers were demanding answers. Within the professional forest community, and even at Collins Pine, where employees knew about the big profit losses to their sustainability programs already, there was enormous resistance to meeting with the certifiers. But their bosses, the Collins family and Jim Quinn, had decided to go green, so a meeting with all the stakeholders was set up.

There was still a lot of hostility between the logging industry and environmentalists back then, another arena greatly improved today. In 1993 Collins Pine's team was headed by Dr. Robert Hrubes, a professional forester and forest economist, but a man "who was known to loggers as 'the guy who presented testimony for the Sierra Club.' " His team's assessment of Collins Pine was conducted through a series of field investigations. "Collins not only passed, but the 1993 executive summary concluded that their scores 'serve as a standard of excellence for other owners of North American mixed conifer timberland to emulate.' "

The Collins Almanor Forest became the very first privately owned forest in the United States to receive this independent certification, meaning the forest's lumber can bear a consumer label stating that it is cut within the renewable limits of that forest. The foresters were enormously mollified. Because certification means inviting an outside agency in to determine if a company's practices are truly sustainable, even good companies fear it, because they worry about loss of control. Collins Pine's former chief forester, Bill Howe, eventually embraced certification, not only because it solidified their markets but because he had to admit it inspired everyone to greater heights of achievement. Terry Collins, the family son running the show in Chester, said, "Collins Pine was getting maybe a little complacent. We thought we knew all there was to know about sustainable forestry. But as we went through the certification process, we found ourselves being asked some challenging questions by a third party. Then we started to realize we could do better. It's really revitalized our practices."

The new knowledge about the key role forests have to play in saving the atmosphere is one more thing that's made Collins Pine run even faster toward their sustainability goals. Terry Collins has become the company's point man on climate change and carbon dioxide, as well as an active member of the California Climate Action Registry program, which is spearheading climate change legislation around the world. Today Collins Pine holds the wwf's Climate Savers award, having reduced their carbon dioxide emissions 15 percent below 1999 levels. Mosby was off to the awards in spring 2009, noting that Collins Pine was one of the seven founding members of the Climate Savers organization, which has since gone international. Its thirty-nine companies include Nike, Johnson & Johnson, Coca-Cola, and Lafarge Concrete. "We're still the smallest," he says proudly, going on to describe the biofilters they put on their motors and their use of surplus biomass (what they don't keep to enrich their soils) to power a cogeneration plant at Chester that produces so much electricity Collins Pine is able to sell the excess to Pacific Gas and Electric Company.

They even made the enormous investment of $13 million in a new sawmill at Chester, which cuts twice the wood but uses half the electricity. "That was a big expense for us," Mosby adds a little ruefully. "We might not have invested so much if we'd known about the current downturn. But even so, we're fine. Lots of lumber companies, much bigger than we are, have cut in half or are going down. Even we have had to lay off some people—we're down to 640 employees from 811 at one forest, and we hate that. But we'll be OK, because we're sustainable. Our forests still have all the timber they ever had! We don't have to extend operations to make money, because our capital keeps growing back. Moving forward, we now have some very valuable carbon credits to sell."

In fact, Almanor is such a functional, old-growth type of healthy carbon sink that it may help us learn why intact tropical forests are getting bigger trees and temperate ones are dying. "We know every tree, how it's doing, and we don't ever clear-cut," says Mosby contentedly. One of the reasons foresters wait years to get a job at Collins Pine is that a major part of their work is walking the forest—monitoring conditions like the temperature of streams, counting the number of osprey, owl, and woodpecker nests, noting whether chipmunks and other animals are moving into the nurse logs that have been left for them—in short, amassing the kind of baseline field data that most biologists and government agencies can only wish they had the time to do. Collins Pine even takes photographs of every single tree in their forest and spends time back in the office getting to know them; it takes two full years to amass each set of photos, and they've been doing it every decade since 1940. This is how they know so much about every one of the trees under their care. They try to set up their cuts so that they function as natural fires would have, doing steady "biomass correction," that is, thinning fire-attracting brush. Except for that, they never cut trees smaller than a foot in diameter and they never, ever cut more wood than they know their forest can replace.

When asked about the pine beetle, spruce budworm, and the other pests that are ravaging nearby forests these days, Collins Pine is not worried. Like any normal forest, these little predators are always around, but Mosby says, with enviable serenity, Collins Pine's practices have kept them under control. "We do single-tree selection in any case. So when it comes to the beetle, we pick out the trees that are infected, take a few around it to isolate the outbreak—and the pests don't spread to the other trees." It's a time- and labor-intensive means of taking care of the forest, paying attention to each tree as a mother does with her children: making sure every individual is doing all right, and if not, mimicking as much as possible the natural fires of the past that sanitized forests by cleaning out the dead and sick trees and killing overgrowths of insects and diseases.

Collins Pine's methods are so careful and labor intensive, however, that very few other forest managers are doing the same, certainly not on the crown lands of British Columbia. There the problem has been compounded by years of mismanagement, fire suppression, and climate-warming trends. In the Canadian western forest, a runaway pine beetle infestation is leading to a mass liquidation of timber that illustrates even more clearly that the philosophic basis for modern forestry is completely at odds with ecosystem reality. Real solutions should not be directed at controlling the pests but controlling the humans: limiting our desires and saving resources for the future. That means allowing nature to take care of itself, by paying attention to natural systems and taking profits that reflect ecosystem realities—not human greed. That's how Collins Pine is managing to laugh at pine beetles, budworm, climate change, *and* economic downturns.

LITTLE BITTY PREDATORS

When harmed by [pests], leaf cells send out signals telling all plant parts to start their defence response... Messages like "water shortage" or "too much shade" circulate from cell to cell within the plant. In the last twenty years it has become increasingly evident that plant cells communicate... in almost the same way as animal and human nerve cells.

Florianne Koechlin, biologist and chemist[9]

It's hard to think of the mountain pine beetle as anything but a serious bad guy, especially if you've seen the interior of British Columbia lately. There, thousands of acres of reddish-brown, not green, mountain slopes spread as far as the eye can see. All those destroyed ecosystems, all the trees and drying streams and crisping moss and lost habitat are due to the pine beetle. These tiny insects were not a real problem until the mid-1990s, when their population exploded. Today the mountain pine beetle has already killed half of the British Columbia interior's pine forest; it looks as though it will have destroyed *80 percent* by 2013. Everyone has noted that the warmer winters brought on by climate change have added to the beetles' numbers. A really cold winter, which last occurred in the early 1980s, used to stop a big infestation dead in its tracks. But along with climate change we have the modern practice of fire suppression to blame for this horrific situation, in which the widespread death of trees is expected to release a billion megatons of carbon dioxide—that's five years' worth of B.C.'s carbon emissions from transportation—into the atmosphere by 2020.[10]

This deadly illustration of the spectacular failure of modern forestry policies shows how mistaken humans are in believing we can understand these complex systems well enough to take over their management. Just as modern managers removed lions from the African veldt and wolves and coyotes from the grasslands of Oregon and Washington, B.C.'s foresters wanted to increase lumber in forests *beyond* what the ecosystems were producing already. They decided the way to get yet more wood was to remove so-called competitors: fire and insect pests. So, like the Oregon pioneers and Allan Savory on the elephant savannas, they embarked on a fifty-year program of ripping predators out of the forest to ensure that humans would have more trees.

The idea that every component of a natural system is in fierce competition for scarce resources lies at the heart of predator suppression. Throughout the last century and a half we've visualized forest growth as a Darwinian contest that results in one tree "winning" and shading or crowding out the other, competing organisms. This assessment inspired twentieth-century managers to very aggressively favor one species over another: if we help one species and eradicate another—favor pines over aspens, elephants over lions, trees over berries—the species we favor will "win."

Fortunately, science keeps expanding, and new research is helping us understand how such approaches are wrong. Many recent studies challenge the idea of relentless competition in forests. Trees, like so many other organisms, do compete with each other for light and nutrients, but they also actively cooperate—sheltering and providing food for many other organisms like fungi and birds. In fact, they even shelter and feed one another. Scientists have recently learned that stands of aspen are united by a shared root system and thus are outgrowths of one larger organism; what happens to one tree affects the entire stand. More remarkably, different tree and plant species aid each other in difficult times. Researchers from around the world studying botanical issues are also discovering that plants so regularly send out distress signals and communicate so many of their needs, problems, and preferences, primarily by releasing odiferous chemicals into the air, that, "there is a constant murmuring in the air, a babbling of smells." These smells are used to attract beneficial insects or warn each other. But more than just smells are being exchanged.

Sophisticated experiments done over four years involving caterpillar feces and vomit, of all things, have demonstrated that when the caterpillar *Spodoptera exigua* (*Hübner*) attacks corn plants to feed on the leaves, a parasitoid wasp species that preys on the caterpillar and controls its numbers "is sure to follow." Researcher Ted Turlings at the University of Neuchatel in Switzerland found that the *Cotesia margini-ventri* is attracted to only those leaves on which *S. exigua* (*Hübner*) have left saliva; when that happens, the plant emits a particular odor "composed of indole and terpenoids... so strong even humans can smell it." In another example of plant communication, tomatoes produce toxins to defend themselves when attacked by caterpillars, along with methyl jasmonate, an odiferous compound used in the manufacture of perfumes, to "warn" other tomato plants that danger is approaching. This response is so crucial that "researchers working in greenhouses are asked not to wear perfume as it could confuse the plants."[11]

Experiments performed by the Ministry of Forests in Kamloops, British Columbia, didn't reveal much about plant communication but did help scientists understand how young trees sprouting in a forest can survive with so little light: their neighbors feed them! Scientists found that birches growing in the sun support neighboring fir trees by feeding them carbon dioxide via mycorrhizae, the underground fungi that amplify the ability of their root systems to draw up water and minerals. The birches are even more generous when the firs are in heavy shade. No one has yet understood what's in it for the birches, photosynthesizing extra hard for another species, or for the fungi that do the transporting. Except that it's looking as if forests are much more than trees. Like the holistic systems analyzed by Allan Savory, they are emerging as a kind of composite organism of mutually evolved, interwoven life-forms that are both competing *and* cooperating.

Suzanne Simard, the B.C. researcher who discovered the underground connection by feeding trees different carbon isotopes, points to the fact that in modern forestry managers often eliminate species like birch—the generous photosynthesizers—because they think they're competing with preferred trees—like shaded Douglas-fir seedlings. She says, "These species we think of as weeds are serving as critical links, and once we sever those links, we affect the stability of those ecosystems."[12] These findings have also strongly suggested to researchers that the berries, mushrooms, wildflowers, mosses, and herbs in a true forest, as well as the microorganisms, spiders, birds, and raccoons and other mammals, are serving their own interwoven purposes in keeping the entire system going. That means, as Collins Pine already seems to have discovered, even if you're just looking to cut timber, a given forest's productivity depends on how well you maintain the diversity of life-forms within it. A really effective forestry strategy seems to require going back to kindergarten and relearning how to share. By the same token, Aesop's fable of the goose that laid the golden egg should be required corporate reading. So long as the whole forest is maintained, we can reap the timber endlessly. But we have been killing the goose.

So now we have a massive infestation of mountain pine beetle, a small forest animal that's forcing us to rethink the survival of forests. This unpleasantly hairy little monster is about the size of a grain of rice and looks a lot like a skin mite. It does appear at first glance to be a ruthless predator, a vicious destroyer of forests that needs to be attacked by modern managers with every weapon at their disposal—fire, cutting, and chemicals. Even Collins Pine uses cutting to control them.

The area devastated by pine beetle in British Columbia is nearly the size of England. It's cost thousands of jobs, thousands of acres of habitat, and $80 billion in lost trees. Worse, the beetle may continue its invasion into the boreal forest to the north, where it could denude the entire Canadian Shield. Once you look more closely into the matter, however, this horrible monster turns out to have a much more complex profile. In fact, the mountain pine beetle is a prime example of an elegant, cooperative life cycle that is interwoven with the best interests of the entire forest ecosystem. It only exists in the first place because of a cooperative relationship with the trees on which it preys.

If a tree is weakened by drought, poor soil, root problems, over-heating, or any number of other stresses, it issues chemical distress signals—odors like those mentioned above. They are picked up by its predators and parasites—the beetles, worms, galls, and other "pests" that upset conventional foresters so much. In a healthy forest the sick tree is inviting the hungry predator to dinner, in much the same way that an aging sheep invites a mountain lion to come and get it. The female mountain pine beetle smells these chemicals and comes calling. There are lots of limitations on her diet, however. First of all, the tree has to be a lodgepole pine, her preferred species, and it also has to be about eighty years old. In the natural forest in which they both evolved, that's a fairly rare commodity. However, thanks to modern management, based on that idea of violent competition, there are now millions of eighty-year-old lodgepole pines (also the logger's preferred species) in the B.C. forest.

For many years foresters have suppressed natural fires, which means the pines haven't been thinned out and more of them have lived longer than they usually do and have sickened accordingly. The human desire for lots of big, fat trees for sawmills has been at odds with the ecosystem's preference for just a few such trees. So the human-managed forest meets all the beetles' reproduction requirements perfectly. They are able to put in tunnels and set up shop, along with some remarkable symbiotic helpers: blue-staining fungi, bacteria, and yeasts that will help kill the tree. These help-ers also protect and feed the young beetles and start the process of decay that will turn the tree's body into new soil for the forest at large. So it's an attack just like that of the mountain lion on the sheep, but it's also evidence of cooperation and balance with the larger forest organism.

As always in nature's balance, it isn't one sided. In a recent article, researcher Arthur Partridge explains, "If the proper circum-stances for the growth of stain fungi are not present, or if the female does not have the correct complement of stain fungi, the tree may overcome the beetle with [a toxic] resin." But the stain fungi, in suf-ficient quantities, can also detoxify the resin. Only if and when the resin is brought under control can the females finally attract mates, build egg galleries by tunneling around the tree, and infect her host's sap with fungal threads. Even if the tree succumbs, while the beetles are working on it they're providing food for other predators and parasites that live in the forest, like "birds, nematodes, bacte-ria, fungi, viruses, and others." Among the many birds that look upon bark beetles as a delightful snack are woodpeckers, creep-ers, nuthatches, and mountain chickadees. But another modern management practice, taking out or burning dead trees and snags (because they "compete for space" with new saplings), means there are few or no large cavity sites for the all-important woodpeckers to live in, and no carpenter ants in the downed logs for them to eat. Fewer woodpeckers means more beetles.

"Ornithologists," says Partridge, "suggest we leave at least thirty-five dead standing trees per hectare [about fourteen per acre] and [an equal number of] downed logs larger than 25 centimeters [10 inches], to provide protection for future stands of pines." It's a little hard to imagine the foresters that work for big-profit multinationals consulting with ornithologists on such refinements. Their goal is board feet for the next quarter, not pine stands for twenty or fifty years from now, and their training has been pretty much the opposite of holistic. Unlike Collins Pine foresters, lovingly studying every tree and measuring every nurse log, the last thing those companies are likely to care about is the well-being of birds. They're worried about their target tree species, international markets, good quality timber, sawmill costs, and labor problems.[13]

Although it sounds as though researchers now know a lot about the pine beetle, they really don't yet understand how their relationship with the smallest beetle predators—the nematodes, viruses, fungi, and bacteria that prey on them—plays out; but they have observed that the insects are often very sick in the latter phases of an infestation. This helps support a predominant theory among many scientists concerning a close relative of the pine beetle, the spruce budworm: that these infestations will naturally burn themselves out, if left alone. Past outbreaks of pine beetle—the last big one started in 1979—either did burn out or were curtailed by a particularly cold winter. Because of global warming, most provincial and commercial forest managers are less confident there will be a really cold winter any time soon, but they're embarking on an aggressive cull of affected trees that may end up, like agricultural biofuels, to be a cure that's worse than the disease.

Because of the fear that all those dead stands will either feed massive fires (the pines are extremely flammable the first few months after they die when they still have their needles) or "go to waste," the big logging companies have been granted permission to salvage the dead wood by doing clear-cuts. However, a belated assessment of what happened to the unsalvaged forests killed by beetles in the early 1980s found that "in the absence of conventional intervention (clearcut logging followed by replanting) these stands had weathered the earlier beetle attack very well. Twenty-five years after the attack, they were filled with both dead and healthy trees of different ages." In fact, "the number of pine trees growing on some sites actually *exceeded* the number of trees forest companies would have been required to replant had such sites been logged."[14]

The current policy of clear-cutting sick or dead forests actually takes one healthy spruce or fir for every two pines. Because the beetles prefer older trees of a particular species, in most forests that still leaves an understory of saplings, brush, and smaller trees that can continue to act as a carbon sink, hold on to the soil, purify water, and provide habitat for all the other forest organisms. The saplings also eventually grow to timber size, creating a usable forest in the next twenty or thirty years. But right now, salvage clear-cutting, continuing well after the needles have dropped and the fire threat is minimized, is the province's and the big logging companies' official response to the outbreak.

A lot of studies are coming out that confirm the B.C. government needs to change its practices, and today they are at least adopting a few suggestions, such as grinding up slash and using it to fuel pulp mills. But the main recommendation coming from science is very simple, one that sustainable forest managers have been repeating for two decades: *ban clear-cut logging in any situation.* Clearcutting and fire suppression are styles of forest management that have proven to be horribly damaging and are easy to abandon; currently they pose even greater threats to forest cover than global warming. Collins Pine's use of careful, selective cutting proves how the right kind of management can weather any storm. Huge multinationals will never manage this way because of their need to satisfy shareholder's demands for constantly increasing profits every single quarter, which include absolute panic when that unsustainable goal isn't achieved. Small-scale, local, adaptive management, guided by international certification standards and regulations from state and federal authorities, is the only way to save these forests, from both pests and climate change.

MORE THAN JUST THE TREES

More than 2 billion people around the world [roughly one-third of the planet's total population] depend on non-timber forest products (NTFPS) for food, shelter, medicine, fuel, and cash income.

Global NTFP Partnership Strategy Document[15]

It's pretty strange that until recently, modern scientific forestry has fostered the idea that forests are just trees and that a monocultured stand of eucalyptus, for example, genetically engineered to express pesticides, is the same thing as an ancient community of beech, brambles, vines, ferns, fungi, flowers, animals, birds, and all the other life-forms and products a complete forest provides. Although traditional forest people, like the tribal peoples of India or the Cree in Quebec, have always been well aware of what they can get inside a forest, the idea of "non-timber forest products" (NTFPS) is only now entering the consciousness of modern managers. In fact, forest clearings, and "edges" in particular, provide, as anthropologist Scott Atran says, "the *highest-yielding, least labor-demanding* source of products that humans have ever known." When forests are appreciated as interrelated, self-sustaining organisms and their rich diversity is allowed to thrive, they can provide thousands of products and benefits to a wide number of species, including humans, almost indefinitely.

Who should manage the continuous, rich production of all these forest products—and who should be able to claim the benefits—has always been an important and contentious question. Beginning in 2005, a new umbrella group, the Global NTFP Partnership, was created to oversee the burgeoning number of academic, private, and charitable NGOs that have sprung up to address the management of all forest production, not just timber. Nearly all of them support what's called "participatory forest management by local groups," which is simply another name for one of the 4cs from the Adirondack Park model—communities that act as "living fences." In southern Ethiopia, for example, the NTFP-Participatory Forest Management (PFM) development project, funded by the European Community, has begun a project to help local farmers in "the improved production of NTFPs, including coffee, honey, bamboo, and spices." It has long been known that coffee, a shade-loving shrub, grows very well within a diverse forest; so do cardamom and many other spices and aromatics. But modern agricultural practices have always tried to isolate species, as part of their belief that intensive monocultures will produce more because they're shielded from predators and competition. They will, but only until the loss of the soil biota, fungi, pollinators, and so-called pests catches up to them, usually in about one human generation, but sometimes after only a few planting seasons. Increased yield in one species, we're learning, does not make up for the loss of all the others. Sustainable production requires, relatively speaking, much simpler technologies and far less extensive interventions. It does demand that the managers pay attention to what kind of ecosystem the product came from in the first place and that they replicate that ecosystem as much as possible.

Sustainable production also requires that the managers and those who profit from the forests are one and the same. As the

NTFP-PFM website puts it, "In order to manage forests in a sustainable manner, it is now recognised that participatory arrangements are needed. These must involve the local communities." The site goes on to elucidate a methodology already in practice in Wemindji, Haida Gwaii, the Periyar Tiger Preserve, and the Amboseli National Park: "Rather than the government 'policing' the forests to protect them... the preferred method now is to recognise the rights of local communities to the forests and their products and give them responsibility for sustainably managing the forest."[16]

Quite a lot of forests around the world are already in private, local hands, and at least some private woodlot owners have learned many of the holistic lessons the forest has to teach. They bought local woodlands years ago and have been making a living from them while trying to make sure they will be passed on to their children and grandchildren. There are many examples of how small- and large-woodlot users, whether they own the land or just have tenure, have learned to be real stewards of the forest. One who's attained near-legendary status is Vancouver Island forester Merv Wilkinson. He's in his late nineties now and retired, and his small, 140-acre holding, called Wildwood, has been acquired by the province as a showplace forest in an area of the island now stripped of its natural ecosystem.

Wilkinson used only selective logging and a philosophy based on one of the primary principles of sustainable forestry: "Establish a rate of cut that sustains the integrity of the forest." Back in 1945 Wilkinson estimated that his forest grew at the rate of 1.9 percent, meaning that as long as his cut never exceeded that, the trees would constantly replenish themselves. In subsequent years the many American, European, private, and public foresters who came knocking at his door showed he wasn't far off; determined by computer, his forest's growth rate was between 2 and 2.1 percent. That means he's taken over one and a half times of his forest's original volume in his long life—and his land still has 10 percent *more wood* on it than when he started. During that time he provided full and partial employment to twenty-six people. The land's generosity, managed this way, could go on forever.[17]

Wilkinson also favored NTFPS like mushrooms, berries, mosses, medicinal plants, and forest wildlife. Like traditional groups and Collins Pine, he didn't need expensive experiments to realize that the nonmarketable species sharing his forest are not competitors for its timber production but indicators of its health and resilience. In an interview a few years ago, he exulted in the fact that he had everything from "four active families of pileated woodpeckers... a pine marten who moves through and cleans out the squirrels, and occasional species of owls," to "otters, deer, beaver, and a cougar." Wilkinson is famous for working so continuously in the forest with these forest denizens that even wise old buck deer don't bother to hide when he's around. He's on the board of Jon Young's Wilderness Awareness School in nearby Washington State, and people there have a joke about Merv's name. They call achieving a relationship with the forest as wonderful as Merv Wilkinson's: "Attaining Mervana."

As we've seen with the symbiosis of the birches and firs, and the mountain pine beetle and its fellow predators, pushing a forest off its natural evolutionary path degrades it more subtly than the sudden crash of a storm or fire, but just as irrevocably. The herbicides laid down to protect sickly industrial clones from the competition of trees like willow, alder, and maple destroy the soil's chance to be rebuilt by their inputs of humus and nitrogen. Alder, long considered a "trash" tree to be killed with herbicides, actually inoculates the soil against disease and pests; that's why nature puts it out first.

And insects, diseases, and pests treat monocultures of trees the same way they treat monocultured food crops. Wilkinson points out: "I've seen disease [and insects] my whole life in the forest, little pockets here and there of all one species. But never, till we got to clear-cutting, did I see epidemics." Even the reassurances we're given about restoring logged areas through tree-planting (including some "carbon-offset" programs) are a joke to anyone who checks on them a few years later. Wilkinson says, "They plant from plugs, a way of growing a lot of seeds in a minimal space. The roots are deformed... there's no proper taproot. The young tree is completely vulnerable to wind... the soil isn't carefully placed around it. If they survive at all, they bush, developing multiple tips. And then," Wilkinson says, "the industry rushes in with the fertilizer. It's like giving a dose of drugs to a drug addict... [It just] kills off all the understory and all the systems in the soil." More fundamentally, as retired forester Barry Ford demands, "Why do they [the big industrial companies] have to plant trees in the first place? What's wrong with their soils and their regeneration? They should be asking themselves if their cuts are too big for the forest to naturally regenerate, the way Collins Pine does."[18]

Plantation trees are not only the same species; they're the same age, and these days, because of the latest cloning methods, they have identical genes as well. Wilkinson tells the story of the massive European blowdown of trees in a 1986 hurricane, in which England lost 14 million trees, Holland 2 million, and Germany 6 million. Wilkinson says, "It hit England first. The Germans knew it was coming, so they sent their foresters out to be where they could observe when the winds came. They were stationed, ready with their notebooks. They had people at the multi-species, multi-aged forests and people in the monocultures. Observers at the mixed-height forests noticed that the trees worked against each other as the wind hit. The primaries, the younger trees, would bend first and the secondaries would bend next, and then run into each other coming back. Results: they diffused the wind, so the loss in those forests was very minimal." Klaus Gros, a senior forester with the German government forest service, told Wilkinson that in such a storm, "You wouldn't lose nearly the trees that a monoculture will. Our... solid mono-height forests, like a field of wheat, they got rocking, and on the fourth or fifth rock, the whole thing went down, like a domino! We have learned a lesson. No more mono-aged forests." And that means no more clear-cuts in Germany.

WHO OWNS THE FOREST?

Biologists and protected area specialists are beginning to change perspectives on human interactions with nature, acknowledging that the traditional management practices of indigenous peoples can be positive for biodiversity conservation and ecosystem maintenance. This positive outcome is best gained by devolving control of forest land to communities.

Andy White and Alejandra Martin[19]

For most of the human past, people used the products of their nearby forests in much the same way they used their gardens and agricultural patches. Today scientific managers are beginning to recognize once more that forests can support many of us *without* being cut down. Examples of traditional agroforestry can be found in remote areas such as the Canadian north, the Amazon, or the Congo, where forests have been carefully managed by humans for thousands of years through burning, planting, and selective cutting favoring certain species and certain kinds of animal habitat.

Traditional agroforestry, unlike its distant cousin industrial agro-forestry, is often so gentle in its interventions that modern scientists can walk through a forest that has been profitably managed by local users for hundreds of years and believe they're all alone with ele-mental nature, or even that the forest has been neglected and needs their management.

Francis Hallé, a French tropical botanist who pioneered the study of canopy life, describes traditional agroforestry best: "Picture a circle," he says. "Within it is primary forest—real jungle, virgin or rain forest inhabited by the myriad creatures that have evolved in such places, from insects and molds through to birds and large mammals. Picture another circle outside it, forming a ring." Within this "secondary forest" are community gardens, often called "swiddens," and on the periphery is the village. From the village at the edge, people walk in to tend their family plots—generally no more than a day's walk inside the secondary forest. They usually have several plots in the same area, but sometimes they're scattered in different directions. These plots are cleared by slash-and-burn, but they're very small, usually less than an acre. They're closely planted with a very large variety of plant species; in other words, they're the typical smallholding polycultures upheld by Indian agricultural analyst Vandana Shiva and by UN food production studies as the prototype of sustainability.

In Guatemala, for example, there will be corn, beans, squash, and root crops such as yam, cassava, and macal. Banana, coconut palms, and other fruit-bearing trees will provide more variety and shade for the crops that need it, like sweet potato. All indigenous groups ensure a harvest by having several small plots per family heavily planted with many species but seeded at different times and located according to soil type and drainage. This allows for staggered harvests and prevents disasters, like animal or pest predation or poor weather, the kind of thing that habitually wipes out farmers who plant monocultures. The plots are tended more or less constantly throughout the season; one of the most difficult chores is guarding them from animals when harvest approaches. Even if one or two plots are destroyed, there are always others.

To supplement these gardens, the people also plant and care for individual trees, shrubs, and other plants. They may have planted them deliberately, gathering seeds or cuttings from the primary forest; or they may have simply fertilized, weeded, and protected the seedlings they found in the area. These palms, nut or fruit trees, and rare medicinal herbs are scattered willy-nilly. They are considered to be the "property" of whoever first planted or started taking care of them, and if, for example, a neighbor should take the nuts from someone else's tree, he would be in as much trouble as a cattle rustler would be from a rancher. The harvest would be physically defended, and the "thief" could be forced to make restitution or suffer humiliation before the elders; he or she might even be fined or beaten.

Most current studies by well-meaning organizations, such as the World Forestry Centre, a branch of the UN Food and Agriculture Organization, are still unaware that most forests are already being "farmed" and "owned" to maximize production. The material coming out of them is supporting large numbers of local people, albeit under a system Westerners don't recognize because it's so different from our own. Fortunately sociologists, anthropologists, and historians have gradually been discovering that under most of the traditional systems used by indigenous peoples, a clear and widespread method of forest production and ownership has evolved—and it's much like the traditional agroforestry that Francis Hallé has described so well. Under this system the forest is a *communal* resource, not a *common* resource, and the garden plots and the valuable trees are strictly private. Only the primary, wild forest is considered a commons; even then, strangers, say visiting from some faraway village, have to ask permission to take anything out of it. A group taking care of particular trees and harvesting their produce within that forest creates "tenure," a convenient, loose term for ownership, in the sense of "holding in trust," while getting use and profit from something. That's the idea of "communal": not everyone can take as they please, but the resource is largely open to the community. The idea of tenure can, but does not necessarily, include written deeds or contracts. Whole groves of timber may be planted or protected in the primary forest and left for fifty years or more, so that succeeding generations of a certain family, clan, or village will have building supplies. They're culled in small numbers limited by tradition and religious proscription. An individual can't drag home many 60-foot logs in a small village without all the neighbors knowing he's taking more than his share, so there's no reason and no opportunity to overexploit. Under this system everyone has a right—within certain limits—to useful material from the primary forest.

Most countries in the world retain some sense of the common value of forested land, and it's very often owned and managed collectively by their governments. Even in the United States, the champion of free enterprise, millions of acres of forestlands are in government hands, and nearly all of Canada's vast boreal and temperate rain forests are held as Crown land. In Germany there are federal, state, and even municipal forests, where everyone can gather dead wood for their fireplaces or hunt rabbits and wild boars, but no one is allowed to take away all the trees or turn the forest into plantations.

The forests in Third World countries are basically held by their governments as well and are left to the devices of the original aboriginal inhabitants until the elites have some need or method of accessing their resources. Then, like forests in Canada and the United States, they can be logged or cleared, because the government will grant "tenure," in the form of paper leases and contracts, to agricultural or recreational developers and large logging corporations. But the underlying assumption remains: forests are held by governments for the benefit of the people. This principle sounds pretty equitable, but of course it hasn't worked out that well in practice, largely because of First Industrial Revolution problems we have already analyzed: overexploitation for industrial goals, a fragmented understanding of a forest's true values, and confusion about the basic philosophy underlying the idea of common ownership.

Modern forest tenure views make liquidation to acquire money the primary purpose of forest ownership. In 1968 professor Garrett Hardin described what he called "the tragedy of the commons": "When resources such as trees are 'free' or open to everyone, costs arising from their use and abuse can be passed on to others. The rational individual has the incentive to take as much as possible before someone else does... Because they belong to everyone, no one protects them. The causes of overpopulation, environmental degradation, resource depletion may be found in [this] freedom and equality."[20]

Traditional users of a forest, typically indigenous but sometimes agricultural peoples who have been in the area for a long time, used to be considered to "own" the resource. But by the 1960s "common" tenure came to be seen as bad for a resource, partly because of Hardin's work. It didn't help that commons ownership was considered to be a feature of communism and socialism, political ideologies the United States and many other Western countries feared and despised. So, the destruction of commons ownership was enthusiastically upheld by governments and forest products industries all over the world for social as much as for practical reasons. It turns out, however, that neither the Western governments and industrialists, nor the communists and socialists, understood how traditional users were really managing their forests.

Many of the accounts in the 1960s and '70s supporting the "tragedy of the commons" theory were written by modern researchers observing forests or other user-managed ecosystems that had already been severely disturbed—that is, the Westerners came in after the traditional rules of the indigenous people had been violated and the locals were adopting outside practices in self-defense (practicing "preemptive" hunting or harvesting, as the Cree of northern Quebec did when their territory was first opened up to sports hunters). Researchers have also interpreted the corrupted practices that people were using from within a partially destroyed social system as the normal way "common land" was always treated, because that's how it was operating by the time an industrial or colonial observer happened upon it.

In fact, when rules that govern common land use are undisturbed, they tend to be more stringent than the rules that apply to "private property." In our society "property rights" often permit owners to despoil their land until it is completely devoid of trees and animals and even to dump toxins in shared watercourses. When an owner's behavior seriously threatens the future of the larger system, the worst thing he or she has to fear is a fine. Traditional violators are fined as well, but they can also be publicly humiliated, be stripped of their tenure rights, be ostracized, or have their herds seized—they may even be beaten or exiled. In West Africa, for example, local people who take more than strictly prescribed amounts of live and dead wood from the common mangrove swamps are threatened with death, disease, and general family ruin by the priests, on behalf of the angered tree spirits. This approach may seem a bit harsh, but in fact the only healthy swamps left in Benin are in areas where people still practice the indigenous religion of Voudoun; people in Christian and Muslim areas have destroyed their mangroves.

Federal or national governments, especially in the First World, are far from the forests they hold in common for their citizens. They are susceptible to the advice of the cadres of scientific experts, not to mention the aggressive lobbies of powerful timber industries. Once they become the managers of their countries' forest commons, governments generally subject these holdings to a series of conflicting management styles, which fluctuate pretty wildly according to each country's economic policies and the reigning theory of proper forestry at the time. That's how Germany, for example, turned all its beautiful, dark forests into tidy and ecologically delicate, even-aged pine monocultures that have been ravaged by windstorms, pests, and pollution.

In the United States and Canada, the horror of losing profits to fire damage, coupled with the growing numbers of people living near forests, led to such enthusiastic suppression of natural fires that Western forests began to lose many of their normal climax species and much of their diversity. Today these forests are clogged with small "trash wood" species, ironically much more prone to devastating fires and prey to pests like the mountain pine beetle. And now another management style is required—controlled burning or salvage logging. The purely ideological horror of sharing a resource with a natural predator, like the pine beetle, has brought a train of tragedy that stems from compartmentalized, competition-based forestry education that grants value to only a few economically desirable species. But when a forest is seen as what it really is—one living organism—and the human users of the forest live within it and *share* its bounty with other organisms as well as with each other, forests retain their natural resilience. Intact forests are able to sequester more carbon and survive the biggest climatic, pest, and economic storms. If we took the time to understand them better and listen to the people who already do, they could be feeding a lot more than just the construction industry.

THE EDIBLE FOREST

We in the past are not felling very much forest. They only plant corn; for this reason, they fell much forest—cedar, mahogany... They are not planting much of anything. For this reason, if it doesn't come out good, this year, the next, they fell more big forest... so now, there is no possibility to have more forest.

Itza Mayan farmer[21]

In Guatemala about fifteen years ago, the young anthropologist Scott Atran taught himself a dying Mayan dialect so that he could interpret the forest science of the people of the Petén Peninsula to the rest of us. Atran asserts that the mode of cultivation practiced by the Itza Maya sustained both the ancient and modern Indian cultures of this region by regenerating their forest's biodiversity—indefinitely. In other words, to an unusual extent, they didn't simply catch or gather the resources of the forest; they used traditional agroforestry to cultivate it as a huge, complex crop. Atran lived with the Itza and interviewed them in their language. As he made comparisons with ancient Mayan crop names, he discovered that useful plants that are considered wild today are in fact the Itza Mayan crops, cultivated *inside* the forest.

Atran points out that outsiders, from the Spanish *conquistadores* right up to the waves of recent settlers from the cities, have been so fixated on the European staples of beef cattle and a few cereals, supplemented by beans, squash, and chilies, that they didn't even notice all the food being hauled out of the forest every day by their Mayan neighbors. It's not by chance that the ramon, or breadnut tree, for example, is Petén's most abundant tree species. Like the sugar maple in northern New England, it was carefully nurtured for generations. Its leaves, fruit, and bark are all edible, and it's nutritionally comparable to corn, except in amino acids, where it's superior. Contemporary Itza confirm that their ancestors used to eat this tree "before corn." They still call the tree an "animal milpa [garden]" and list thirty species, including howler and spider monkeys, toucans and macaws, that depend on it for survival. The brush around established ramon trees is controlled with machetes to help seedlings and fruit production, but sparingly, because the Itza believe that "too much thinning allows in too much sunlight and wind, which causes the protective vines and mosses on the tree to dry, the bark to split, and the trees to die."

Far from understanding the uses of the mature forest around them, the early Spanish settlers complained, during their frequent periods of cereal crop failure, that they were "forced" to eat ramon, sweet potato, manioc, yams, plantains, and mamey sapote fruit. The fact that these foods were always available during so-called famines shows that they were suited to the soil and climate, unlike the Spanish exotics, and moreover that the locals were still nurturing them for their own use and had enough to share. Like all agroforestry practitioners, the Itza Maya had small plots scattered around the secondary forest, several per family, with varying soil type, crop mix, and planting times. They didn't experience the Spaniards' devastating crop failures. The species mix made for a delightfully diverse diet of everything from fruit and starch to flavorings: wild pineapple, monkey-apple, mamey sapote, and allspice. As well, the Itza benefited from the forest's production of cedar, mahogany, logwood, locust, coyol palm, cabbage palms, chicle, and uncounted medicinals. This was in the tropics, of course, which can produce food species far more luxuriantly than a temperate forest. And since it's in the tropics that people today are having the most trouble getting food, it's worth knowing that forests can provide crops and fill human physical and economic needs—without being cut down.

Atran calls the local garden plots "artificial rain forests" because they're incredibly high yielding and require little labor. He says they imitate a mature forest's diversity and preserve its nutrient cycle while supporting species deemed most useful to people. This is not the slash-and-burn agriculture, also considered traditional, that has decimated the Amazon rain forest but a more ancient, indigenous version of traditional agriculture growing out of an intact culture. Trees are not fully cleared but are thinned to provide shade where needed. Weeds are suppressed by root and vine crops like manioc and squash, rather than being pulled. Crop rotation restrains the effects of species-specific pests, drought, or too much rain. These gardens have supported hundreds of thousands of people without being changed or expanded. Even allowing for fallow periods and occasional crop failures, they can produce a remarkable 800 pounds per acre of food per year, which is comparable with the yields of many modern food crops. Most importantly, linguistic and archeological evidence reveals that these practices have been fully sustainable for at least *four hundred years*, because lists of Mayan food sources and the remains in middens perfectly match the composition of surviving Itza Mayan gardens.

Atran and several of his colleagues at the Center for Cognitive Studies, when working at the Bio-Itza Reserve, San José–Petén, Guatemala, analyzed the difference between the Itza care of the forest and that of the more recent immigrant ladinos from the cities. The newcomers use modern methods of large, monocultured fields that are all corn, all beans, or all tomatoes. In this lush, tropical ecosystem there are a great many pests, from molds to viruses, and a monoculture does not survive long. The use of pesticides, fungicides, and other chemicals to keep a foreign farming style going degrades the land so quickly that the new farmers must continuously cut more forest to get at fertile soils. But their practices dry the soils out and degrade them, so, as the Itza have noticed, trees can't come back very easily.

The Itza Maya, like most traditionalists, view the environment and all its species, including people, as profoundly interdependent—as part of a social, cultural, and spiritual whole. The remaining Itza have seen much of their fertile, edible home disappear. Today they say about their forest, "Listen for the sound of the jaguar. When there are no more jaguars, there will be no more forest. And then there will be no more Maya." Considering what science has learned in the past few years about the holistic nature of forest ecosystems, the interdependence of trees, fungi, pine beetles, birds, microorganisms, ferns, flowers, insects, and reptiles, right up to the largest keystone predators like jaguars and humans, it's obvious that the Itza are not just being poetic. They are forest people; they know that without its top carnivore, the forest and their culture will not survive. Today we are finally learning enough about these interrelationships to pay more attention to the practices of the remaining forest users like the Itza. Modern managers, through such initiatives as the new NTFP programs and non-timber certification standards, are finally starting to understand what a forest really is and how it can be both used and preserved. If this nascent movement spreads, it could translate very rapidly into a rebirth of healthy forests worldwide.

TURNING POINT: THE ADIRONDACK PARK GOES GLOBAL

Ecosystems behave in complex and sometimes surprising ways. We don't always know the effects of development or management. *Adaptive management* provides a way to decide what management actions can be done safely, what actions need to be monitored, and what actions should be avoided until more studies are done.

Coastal First Nations Turning Point Initiative Learning Forum[22]

"Adaptive management" is a term that helps Western-trained scientists and foresters understand how local and traditional groups manage their resource needs. How did native North Americans figure out that sharing the trees with fire increased forest productivity? How did the Itza Maya learn which food plants would grow in the forest? How did three generations of Collins Pine managers learn to keep their timber safe from pine beetles? These indigenous and local managers all learned by acting and observing over a relatively long period. But since modern managers don't like to pay attention to traditional beliefs, cultures, and religions—which are generally inseparable from traditional scientific and management systems—this modern term has come to everyone's rescue. Today, as anthropologist Fikret Berkes puts it, "Traditional management can be reinterpreted as adaptive management. Alternatively, adaptive management can be considered a rediscovery of traditional management." An alliance of coastal First Nations in British

Columbia is also fond of this term, which they simply define as "a way to learn from management *actions*."[23]

The Coastal First Nations of British Columbia—an alliance that includes the Wuikinuxv, Heiltsuk, Kitasoo/Xaixais, Holmalco, Gitga'at, Haisla, Metlakatla, Old Massett Village Council, Skidegate Band Council and the Council of the Haida—is using this term to explain their involvement in the largest and most promising forest conservation project the southern part of Canada has ever seen. An area the size of Ireland (14 million acres) is going to be managed by multiple stakeholders for multiple purposes. As in the Adirondack Park system, there is a core: 5 million acres protected from logging, as well as an additional 1.7 million acres of fully protected old-growth habitat. The logging that is allowed will largely be eco-system based, and the local users and residents that have been there the longest, the native people, will be mostly in charge. As of March 2009, this project became a fully fledged reality, with $120 million in private and government funding already filtering into coastal communities.

Today's huge, complex agreement began in complete disarray, with tourism, hunting, and fishing operations fighting for tenure rights with loggers and real estate developers. The many small native groups who were the original owners of the area, who had managed it so sustainably for thousands of years that there was still something to fight over, had little status and were often not speaking to each other. And then there were the various arms of municipal, provincial, and federal governments, all vying for rights and eschewing responsibilities, along with the "tree-hugging" environmental NGOs and the general public. It looked hopeless.

And yet, an accord was reached. Today the area covered by this agreement extends from the northern tip of Vancouver Island all the way up to the Alaskan panhandle. This narrow strip of forest pinched between the Pacific Ocean and the coastal mountain ranges represents *a quarter* of the planet's last temperate rain forests. It's eminently worth fighting for, as it supports the highest biomass—the weight of living things per acre—of any ecosystem on the entire planet! It's an awe-inspiring place, where Sitka spruce, hemlock, western red cedar, and Douglas-fir tower as much as 325 feet into the sky. This ecosystem is home to many of the world's last healthy salmon runs, as well as pods of orca, beds of fat oysters, flocks of eagles, starfish in a bewildering array of colors and shapes, and the magical, pure white Kermode or "spirit" bears of the coastal islands' Great Bear Rainforest.

Temperate rain forests are exceedingly rare ecosystems, occupying less than one-fifth of 1 percent of all the land on Earth. Most of them are already gone, and much of this one has already been razed. It's the same incredibly rich forest ecosystem where most of Collins Pine operates, and it's also home to Merv Wilkinson's Wildwood. These forests provide high-quality lumber and paper products, but as the most valuable stands of fine-grained old-growth cedar and spruce have been removed, pressure on what's left has been causing collapse in its key components. A water-dwelling carnivore, the Pacific salmon, is so pivotal to the health of the temperate rain forest that the ecosystem is also sometimes called the "salmon forest."

How did such an enormous amount of rich ecosystem, claimed by so many interest groups, manage to become protected? Tara Cullis, who helped establish the Turning Point Initiative, says it came down to one thing: "How much the native groups were ready to give up." In short, most of the heavily protected areas will come out of native holdings, not those of logging companies or municipalities. Native groups made this sacrifice to save the salmon. For many years environmental and First Nations groups have been warning that logging companies, abetted by government policy, were causing salmon populations to collapse by overcutting the forests. Company officials countered by claiming that overfishing and ocean warming were the real problems and that their industry only wanted to cut 7 percent of the territory—the area around the river banks. However, riverbanks and lakeshores are the engines of the ecosystem. For the past few years the logging industry had been poised to log half of all the remaining old-growth temperate rain forest left in North America—this mid-north coast of B.C.

Most levels of government consider logging interests to be major stakeholders because of the jobs they provide. But the people with the clearest claim to legal ownership of the north coast of British Columbia are the tribal groups—the people of the salmon forests, who never signed treaties with the Canadian government. Numerous court cases have determined that these tribes still have rights, not only to govern themselves but also to fully own large areas of provincial land. Canada's Delgamuukw Supreme Court ruling of 1997 in particular established that aboriginal title still exists in this area and cannot be ignored.

However, over the last century British Columbia's settlers have also claimed its riches: the salmon fishers up and down the coast who depend on the fishery for their income; the truckers and sawmill workers who live off the forest industry; all the town governments that have sprung up to serve the people living along the coast; and not least the tourism operators and various other nature and industry groups. These "resource stakeholders" also demanded a seat at the negotiating table along with native groups, the federal and provincial governments, and the forest industry.

Even as legal and administrative battles were being fought about who owned what, the fisheries continued to be overharvested and the forests clear-cut. Major environmental organizations, notably

Greenpeace, the Sierra Club of B.C., and the Rainforest Action Network, launched a national campaign in 1997, demanding that a chain of north coast watersheds be put aside as parks. They took the Kermode bear as their symbol and dubbed the region the Great Bear Rainforest. Thanks to the relatively new tool of timber certification through the Forest Stewardship Council, they were able to launch an international boycott against wood extracted from B.C. forests, and the Great Bear campaign took off.

The companies and the government were forced to the table by this public pressure and eventually agreed on an eighteen-month moratorium on logging in a hundred watersheds until final preservation status could be determined. Of course, it wasn't just the boycott that brought down the logging industry. The timber companies involved realized that much of the forest that was economically accessible had already been liquidated. But they could also see that, in both Europe and North America, consumers and mills were demanding more certified wood and that even wild animal protection was becoming a critical issue, as initiatives to preserve vast corridors of wilderness gained momentum. In short, the industry was being forced to change its ways if it wanted to keep cutting at all. It had to start paying closer attention to the environmentalists' demands in order to regain lost markets and public credibility.

While the international campaign to boycott B.C. temperate forest lumber and pulp was gaining traction, the David Suzuki Foundation, a very interested observer, adopted another position. The native groups' clear statement that they needed to have the right to exploit the economic potential of their lands, to log and fish on whatever land they ended up getting, at first stood in the way of making alliances with most environmental groups. But the late Jim Fulton, a former member of Parliament for the region, was head of the DSF at the time and led the Foundation to recognize the original inhabitants and stewards of the central and north coast in their claims. These groups had proven they could support rich, diverse cultures and the highest nonagrarian aboriginal populations on the planet for millennia without destabilizing the region's productivity. Fulton realized that these people had both exploited and maintained their massive salmon runs, rich ocean resources, and productive forests, and decided that was a good enough reason to want them in charge of managing the territory into the future. Of course, in the old days, the native groups had hand axes and simple fishing tools. In order to address the modern threats of industrial fishing, logging, and clear-cutting, the DSF and their new native allies understood that once the bands' claims over the territory were recognized, they would have to follow traditional cultural and economic practices, living in the forests and managing them for the future, not by setting aside parks or preserves but not by exploiting them in the usual industrial way, either.

The *quid pro quo* for the DSF was that ecosystem-based techniques would be used to manage the forest. Fulton sent Tara Cullis into all eleven of the coastal First Nations communities to establish a formal working relationship. For two years Cullis worked closely within these communities to learn their needs and concerns firsthand and also to assess their strengths and capabilities. This new relationship led to the formation of the Turning Point Initiative, an alliance of the native groups, many of which had contentious rivalries over disputed lands. Their partnership eventually gave the First Nations communities the strength to gain formal preservation status for their land claims in 2009.

The $120 million that is the centerpiece of the Great Bear Rainforest deal today is helping these small, scattered coastal groups—who suffer from as much as 90 percent unemployment and the subsequent loss of their children to the cities—support themselves sustainably. Half this money will fund "sustainable economic diversification and infrastructure," such as viable shellfish and seaweed harvesting, tourism, green energy infrastructure, education, and sustainable logging and fishing. The other half will go into an endowment "to support conservation management and related job creation," which should provide steady employment for the original managers to keep doing what they're so good at. Like Haida Gwaii's Watchmen—the stewards of Gwaii Haanas—they will be the ecosystem restorers, conservation managers, and assessment officers throughout the 15.8 million acres of the Great Bear Rainforest.

What is extremely reassuring to see, even if it's not surprising, is that like all the local management groups discussed in this chapter, the coastal First Nations are concerned with every single aspect of the ecosystem, and they quickly set up three land-use objectives. These objectives cover not just conservation but also First Nations heritage and culture, which overlap. For example, cultural tradition mandates the protection of several key tree species, including red, monumental, and yellow cedar, which are to be spared even when found within logging cutblocks. The coastal nations' "aquatic objectives" are to "maintain water quality and quantity as well as ecological function in riparian forests, swamps, wetlands" and coastal areas, eventually seen as extending 6 miles out into the oceans. That means the coastal nations will monitor not just their favorite species, the salmon, but tailed frogs and salamanders, northern goshawks, marbled murrelets, and black-tailed deer. Their "biodiversity objectives" center culturally on grizzly bear habitat; like the Itza Maya, the coastal nations know that without the top predator, the forest will disappear.[24] The agreement also provides for "networks of reserves" to connect the current polka-dots of old-growth forest habitat, eventually linking it to a future corridor stretching from the Yukon to Yellowstone. Sustainable fisheries and tourism, as well as green buildings and non-timber forest products, especially medicinals, will be supported with the new funding; net-cage salmon farming, trophy hunting, mining or drilling, and unsustainable logging practices will be left unfunded and, in most of the area, prohibited.

Possibly the most remarkable part of this whole preservation arrangement is summed up in a *Vancouver Sun* headline that appeared in April 2009: "Great Bear forest deal shifts power after years of grinding negotiation."[25] Indeed, there has been the unlikeliest power shift imaginable. Power has moved from obscenely rich and powerful multinational corporations and their staunch government allies to a handful of tribal people, whose villages sometimes number under a hundred souls, who have no money and little influence on the larger society. Cullis says that the six-member steering committee for the new marine component of the Great Bear preservation area has one representative each from the federal Department of Fisheries and Oceans and the provincial department of the environment—and four more members, all from First Nations communities.

She chuckles, "The deep-pocket industries like the major logging companies, industrialists, tanker port operators, fishing lodge owners, truckers, mayors, and all that attending this meeting were very shocked. About twenty minutes before the end they wanted to know, 'How come there are so many natives holding seats on this powerful committee? I mean, four out of six!' Of course, natives are normally never let into these discussions at all, so no wonder the industries were shocked. It's partly because they were wearing different hats; Art Sterritt was there as director of Turning Point, and two representatives were from the central coast plus one more was from Haida Gwaii—but they also all happened to be Turning Point people. If we succeed, they'll have control over that whole, huge marine area, as well as the forests. And the native people feel, 'Well, why not? We're the reason it's all still there in the first place!' I'll tell you, this whole situation is giving nonnatives a huge education on racism and society's normal assumptions about who should be in charge, and they started getting lessons that day."

Because these once-quarreling native groups were able to present a united front, they were able to work with the environmental NGOs they first distrusted. And because that cooperation meant some hope of legal tenure over the ecosystem, the NGOs were able to help raise $60 million from private foundations and pressure governments for the rest. All that funding, says Cullis, "has been divided into different buckets, some for communities, some to manage the forest and the rivers." Cullis is the cofounder and president of the David Suzuki Foundation, which got the native groups to unite and work with the environmental NGOs through Turning Point in the first place. The DSF is no longer so closely involved with the Great Bear forest because the level of ecosystem protection isn't high enough. Cullis says, "They lowered the bar so they could raise all that money. Still, even though forestry standards aren't as high as we wanted, they make it *so* much better than it was!"

It's all reminiscent of what happened in Adirondack Park. Everyone's working together, but they all complain and each group wants more. It's a complex, flexible, protean, community venture, led and implemented by locals, with immovable government protections over a wide variety of ecosystems and uses, maintained through largely self-supporting funding. This is the model that has been proven to work over time and over increasing tracts of the planet's remaining forests, from Periyar in India to New York State. And 150 years from now, it's likely to be running along just as fractiously and—from the point of view of the bears and murrelets— just as well.

STEP ONE, CHECK; NOW, STEP TWO

Europe's proposed new timber regulations have been transformed from a ramshackle statement of intent to a credible framework for controlling the illegal trade in timber.

News release from the Global Forest & Trade Network[26]

A couple of good companies here, some well-managed private or community forests there, as well as increasing numbers of very large new stretches of parks and preserves can be found scattered around the world. But if we're going to save our water, air, and soil, we need to protect a whole lot more of our forests—and do it a whole lot faster. The loss of forests means the loss of soil, water supplies, atmosphere, climate, food, and human culture. We now know that forests function as a primary bulwark between us and runaway global warming. For this reason alone, we cannot allow them to be liquidated for a few short-term economic benefits for a very few already wealthy people. A lot of important people and organizations seem to be finally getting that message. Because it implies serious transfers of wealth and control over resources, however, it's a message that has to be understood and fully supported by people in every walk of life, including those who have never set foot in a forest.

Urban citizens need to get busy on this one. A recent study by the World Bank, together with the WWF Alliance for Forest Conservation, established that protecting forest areas is the most cost-effective way to supply most of the world's biggest cities with safe drinking water, which "provides significant health and economic benefits to urban populations." More than a third of the world's largest cities—New York and Los Angeles, Tokyo and Melbourne, Barcelona and Nairobi, Mumbai and Rio—rely on protected forests in nearby catchment areas for most of their drinking water. Without those forests, they will run out. It's been proven that protecting forest watersheds can be seven times cheaper than building and operating a water treatment plant, another advantage on top of all the other benefits the protected forest confers.

All over the world people think that watershed, forest, and agriculture issues are "rural problems" and have nothing to do with city dwellers. But the glory of Istanbul could be destroyed by water-borne disease if illegal housing developments, local disputes, and inappropriate land-use policies keep eroding its watershed. Mount Kenya, which bestows a bounty of snowmelt on Nairobi, is increasingly beset with illegal charcoal burning, logging, and road construction, and the water supply to the city is being severely affected. New York discovered years ago that keeping their watershed pure has enabled them to avoid water treatment costs. If such forests are managed properly, the way it's happening in B.C., the people living near them could find themselves prospering and looking forward to safer tenure of their homes.[27]

Today we've come to recognize in detail just where First Industrial Revolution practices have been going wrong. Researchers are gaining a very clear understanding of what sustainable forests are, and what this vitally important and beautiful ecosystem needs to survive. Management collectives from the village to the international level are combining and applying Second Industrial Revolution knowledge—like TNS principles as practiced by Collins Pine and adaptive management as used in Wemindji and the Great Bear Rainforest, including the Adirondack Park model—to huge swaths of protected areas all over the world.

The Ngiri-Tumba-Maindombe area around Lake Tumba in the Democratic Republic of Congo, more than 16 million acres—or twice the size of Belgium—was placed under protection along these lines in 2008. Tumba is threatened by oil palm plantations, commercial farming, and city development, and along with its 150 species of fish and three kinds of crocodile, a lot of humans rely on the area. One reason it has gained UNESCO Ramsar Convention status is that wetlands, along with forests, serve as major carbon sinks for the world, absorbing carbon from the atmosphere and storing it.

We need to remember, though, that putting an area under state, national, or international protection doesn't mean it's safe. Shocking numbers of forest ecosystems are not actively protected even after receiving recognition as critically important areas. Whether named a UN World Heritage site, like Gwaii Haanas, or granted Ramsar status, like Ngiri-Tumba-Maindombe, the forests are protected not by international proclamations but by local people's on-the-ground commitment, with the support of their governments. Without that, all the international awards in the world won't preserve them. In Cameroon, for example, the UNESCO World Heritage site of Dja is characterized by that agency as "one of the largest and best protected humid forests in Africa." About the size of Wales, it shelters elephants, gorillas, chimpanzees, and forest buffalo. But another source mentions that "Dja... harbours 100 permanent hunting camps supplying bushmeat to 128 villages on its perimeter." That's 3 tons of bushmeat—former elephants and monkeys and gorillas—being harvested every single day.[28]

In fact UNESCO status does not come with a management plan or the funding to get one. UNESCO doesn't have a lot of power and status within the UN to begin with, even though it does try to help; around $300,000 has been donated to Dja, "chiefly to train the meager park staff." But there's no UNESCO representative in Cameroon, and no one from the organization has even visited in the past three years. This example proves once again that you can't manage an ecosystem from afar. Appeals to the UN from the Congo's ministry of environment and forests for more money have been met with threats to remove the prestigious World Heritage status. The guards have little power, and the meat they confiscate is auctioned off to feed state coffers, yet another opening for abuses. One heroic Dja eco-guard died in prison after trying to stop a poacher who was supplying an elite client; another was savagely beaten for the same kind of thing.[29] This situation could be fixed, but UNESCO would have to get far more involved than it ever has been in the past, working as the Supreme Court in Canada did to broker a deal between stakeholders, governments, and local people. Regulations could be applied in a country like Cameroon if the right incentives were in place: trade deals, market pressures, and especially aid tied to forest protection. In the meantime our lists of places that have been "saved" must be carefully monitored.

The Great Bear Rainforest can now serve as a model for stakeholder and management action that leads to serious long-term protection. An effective alliance emerged to protect the region because a federal government recognized tenure rights. This enabled the stakeholders to commit to multi-use ecosystem management, based on the Adirondack model and adaptive management techniques. They were helped by socially active NGOs that were able to influence industries through boycotts. They were also supported by infusions of money for local management from government and private donors. Haida Gwaii, in particular, is a showpiece for how traditional managers can form alliances with local towns, loggers, and fishers to make protection work for everyone's benefit. It's vital that such models become better known, because these same mechanisms could be applied in Dja and other places like it. Periyar, in poverty-stricken Kerala, India, was able to direct its own transformation, so hopefully the international community doesn't always need to get involved. But in some countries, it needs to. Fortunately, more tools to support this transformation are becoming available all the time.

Forest certification standards have been international for some time, but now they're getting teeth. The European Union's new "traceable timber amendment" will enable all buyers to track wood and make sure it's not being harvested unsustainably. The WWF estimates that between 15 and 20 percent of European wood imports as recently as 2006 came from illegal sources, "with much of this timber coming through Russia and Finland." EU members will now be able to trace timber back to its source and make sure sustainable agreements are being enforced.[30]

The real key is international trade financing mechanisms that favor the preservation of *all* our remaining forests. If international trade regimes make it difficult for Indonesia to sell its palm oil but offer it big money for carbon credits, it will stop liquidating its forests; ditto for Brazil and Cameroon. We don't have much time to do this, but it does constitute a flush of pink on the horizon. In March 2009, the first summit between the environmental community and leaders in the financial services industry was held in Washington to "explore how sustainable investing and financing mechanisms can help advance the conservation of the world's forests." The summit concentrated on tying "responsible lending" to countries that control the use of their forests and allow proper certification standards.

A whole new global trade mechanism to make sure aid is tied to good practices, called the Finance Alliance for Sustainable Trade (FAST), has joined the highly active Global Forest & Trade Network (GFTN) of the WWF and is using the World Bank, the International Finance Corporation, and the Citi Foundation to set up responsible lending "that adequately assesses environmental and social risks, especially for sectors like the forest industry." Providing better access to business finance for small and medium-sized enterprises in Third World countries means these countries can start developing alternatives to liquidating their natural resources. In this way FAST hopes to "play a cutting-edge role in the promotion of a new, green economy."[31]

Results are already beginning to roll in. In July 2008 the WWF announced that 2.5 million acres of Congo Basin forests "have achieved certification under the world's leading sustainable forestry scheme"—that is, the same Forest Stewardship Council that was a big part of saving the Great Bear Rainforest. Because "illegal forest exploitation and forest crimes are largely due to poor governance and insufficient law enforcement," says Laurent Somé of the WWF Central Africa Regional Programme Office, sustainable management practices provide double dividends well beyond the conservation of vital rain forests. Criminal activities are reduced, and communities benefit from controlling resources in their area.[32] And there's more.

Many environmentalists have despaired when it comes to China. There, a combination of relentless population pressures and a government that is nondemocratic, secretive, and obsessed with financial gain has made most environmental initiatives nonstarters. But the trade and certification approach is bearing fruit even in China. In early 2009 the WWF was able to announce that "forests owned by members of the Chinese chapter of the WWF's GFTN and certified by the Forest Stewardship Council went beyond a million hectares [2.5 million acres] for the first time." Preserving south China's forests is a herculean task, complicated by a hungry market for abundant types of desirable timber, complex geographic features, and chaotic forest tenure systems. It turns out that a huge multinational, Fujian Yongan Forestry Group, has gotten behind certification of its 240,000 acres so that it can reach a larger market. Another GFTN-China member, the Muling Forest Bureau, located in a prime conservation priority area, was supported by the retail giant IKEA in the certification process. It bumped certified acreage up by another 640,000 acres.

Even Russia, classified by Greenpeace as in a real mess, has just adopted "a unified information system" for keeping track of illegal logging in the country's gigantic boreal forest, one of the world's major carbon sinks. With China, the EU, and Africa coming on board, it's obviously getting harder to be the outlaw and still get to market. The U.S., under George W. Bush, also enacted a prohibition on trade in illegally logged wood products, in October 2008. All plant material sold has to declare its origin so that it can be traced. "Illegally logged wood...has been shown to support organized crime around the world," so most communities will be glad to hear the market for this kind of exploitation is being eliminated.[33]

It's exciting to realize that every one of us can help preserve forests all over the world by buying only certified wood whenever we fix up a bathroom or build a fence. Besides making sure that the trees we do cut down are sustainably managed within our own countries, we can work to make sure that forests are valued for *all* their products and especially for their function as carbon sinks and biodiversity engines. Internationally, we really have to get behind a new kind of foreign aid. As one study states, "rich polluting countries should be transferring substantial resources to countries with tropical forests to reduce deforestation rates and promote alternative development pathways."[34] This method of sharing our sunshine with countries laboring in the shade will save not just the forests, but the air, the water, the plants, the animals, and all of us as well.

5

A RIVER RUNS THROUGH IT

FRESH WATER

FRESH WATER HEROINE

Water shortage is as much a source of climate change as green-house gas emissions. We've missed this connection to global warming. We paved the water-retentive land in cities, we've rerouted, dammed, polluted, and wasted—we've prevented the hydrological cycle. And without one, this planet doesn't work.

Maude Barlow[1]

Maude Barlow is known the world over as the premiere activist-expert on this planet's hydrological cycle and how it needs to be managed. In the past she was always characterized as "a voice of dissent" or "Canada's Ralph Nader," a person without access to the politicians running the show. As recently as 2001, in a CBC film biography, she was portrayed by critics such as Tom d'Aquino, then of the Business Council on National Issues, as "someone out of step, who is not able to win public support for her ideas in the demo-cratic court."[2] Although he still has plenty of influence in Canada, Tom D'Aquino must have been discomfited to see the person he tried so hard to dismiss placed in a new UN post that was *created for her*. Only a couple of years after winning the Right Livelihood Award, also called the "Alternative Nobel Prize," with *Blue Gold* coauthor Tony Clarke, Barlow was appointed the United Nations' first Senior Advisor on Water, working with the president of the UN General Assembly, Miguel d'Escoto Brockmann, which means that for the year 2008–09, she had the ear of what can arguably be called the most important democratic forum in the world. When Barlow alights from her many international trips she also runs one of the biggest NGOs in this country, the Council of Canadians. It was founded to protect Canadian institutions like public health care. It was through her work with the Council of Canadians that she first became alerted to the problems the world is having with fresh water.

Canadians have been raised to believe that our country is awash in water—statistics like "20 percent of the world's supply" have been bandied about for years, and if you fly across the northern reaches of almost any province, you certainly will see as much water as land. All those lakes and rivers have led Canadians to be smug about this resource. In reality, however, as aquatic scientist John Sprague notes, this overestimation of our riches comes from including the volume of the water sitting in Canada's lakes. They would have to be drained for the water to become useful to people, and that would destroy entire watersheds. A closer estimate of Canada's share of the world's fresh water would be about 6.5 percent; that's our share of *renewable* water, the part of the cycle that falls from the sky and merely passes through lakes and rivers as it heads to the sea.

It's so important to remember that without water, there is no life as we know it. Humans, for example, are more than 60 percent water. If we visualize the thin skin of life stretched over the globe as a kind of superorganism, we can see that it, too, has water running through it—the rivers and lakes are arteries and veins; the oceans and the hydrologic cycle operate as its heart. In fact, like the beating of our hearts, water is so much a part of our surroundings and our bodies that we usually take it for granted. After all, water simply falls out of the sky; it's all over the ground as well, in streams, ponds, rivers, and bogs, pouring out of millions of taps in every big city, all over the world. People who live in northern Africa, India, or parts of China and South America may not feel so complacent, but even there it would come as a surprise just how little fresh water is actually available for use on this planet. Less than one half of 1 percent of all the water on the planet can be used by non-oceanic life; the rest is seawater or is frozen in polar ice caps.

The only way we can normally get at this usable fresh water is from rain, which today is called "green water" because it's a truly renewable resource. Rain can be caught in reservoirs, lakes, rivers, aquifers, and permeable soil, but to be any use to us, it has to fall on landmasses, not out on the oceans. Millions of gallons of water seep just below the surface where people can access it by digging shallow wells; this groundwater is termed "blue water." More water collects and is stored over time much deeper in aquifers, mysterious underground rivers, and huge hollows in the bedrock deep within the Earth that are accessible only by drilling. That's "red water," least renewable of the lot, and now there's also "black water," which is the sewage and chemically contaminated wastes of our life processes and especially of our modern industries. It's possible to reclaim black water and even turn it into green water again, but only with time, large investments, and a serious understanding of what we're doing.

For millennia, traditional methods of obtaining green water included collecting rain in small barrels and cisterns or huge tanks, building artificial lakes, and trapping water in catchment systems like terraces or small dammed areas and ingenious underground canals. These historic technologies were largely limited to catching rain from above, rather than pulling water up from below. As we hauled the green water out of shallow wells in buckets, more rainwater percolated down, constantly replenishing the supply. Our ability to obtain the water that's stored hundreds of feet below the surface in deep aquifers, however, is very recent. Some of this red water is probably renewable, but much undoubtedly is not; we still don't understand much about underground aquifers. Really deep wells that exploit huge underground aquifers like the Ogallala Aquifer, which stretches across nearly the entire American West, began to be made only around the middle of the twentieth century, when gasoline-powered drills that can pierce rock were developed. As this new technology spread, making deserts bloom and eliminating a lot of physical labor, most of the world's old cisterns and small dams fell into disuse.

Today, with gas-powered machines, huge amounts of water can be pumped great distances and trapped behind ever-larger dams for use in distant locations; in fact entire river systems are now habitually diverted so that their contents not only irrigate areas hundreds of miles away but are harnessed for electrical power while doing so. The California desert, Israel, Jordan, and many other arid lands have suddenly, especially since the 1950s and '60s, turned green with lush crops. Big dams also provide hydropower for cities and for pumping yet more red water out of aquifers, greatly increasing the ability of humans to control nearly all of the planet's lifeblood.

But there's a problem here as well. As anyone who understands the hydrological cycle already knows, of all the natural systems on the planet, water is the most obviously finite. It's exactly the same pool of water that goes around and around the globe, again and again... and again. Drops that quenched a dinosaur's thirst and bathed Julius Caesar might be sitting in your morning coffee. Most of the water raining down and being sucked back up again by evaporation, only to rain down once more, stays in the water cycle. But some of it has been stored for very long periods in those deep underground aquifers. Because of our growing numbers, modern industrial and agricultural practices, and new damming and drilling technologies, this water has also entered the above-ground cycle—with what ecosystem effects we have no idea. Much of it is also being polluted or wasted. There has been so much drilling, damming, waste, and contamination, in fact, that water experts around the world agree that we're now facing a very scary crisis, no less dangerous than what we've done by pumping up fossil fuels from below the Earth's crust and putting them on the surface and into the atmosphere.

And while there are many sustainable solutions, one of the most common suggestions, desalinating sea water, which is often presented as an obvious way to gather immense amounts of water for human use, is not one of them. It has to be done along shorelines, the richest habitats and nurseries for marine life in the world, and the desalinization process leaves mountains of poisonous salt that can leach into the water and permanently destroy these absolutely vital areas.

Maude Barlow has made it her unenviable job to warn the whole world of the unfolding water story and the problems we're going to have if we keep taking this resource for granted. "The water crisis is only about five years behind the climate change one," she informs her increasingly sober audiences. "There is increasing evidence that the warming of the planet is partly *caused* not just by CO_2 emissions but by the way we are interrupting and impoverishing the normal hydrological cycle." She's pitiless in listing her version of "inconvenient truths" that are even more upsetting than Al Gore's. "Australia is burning. Israel is in its worst water crisis in eighty years... Kenya is having a national water emergency; southeastern Spain is desertifying; in Pakistan the glaciers that feed the Indus, their only source of water, are melting and evaporating. In southeast Asia, half the region's supplies are threatened." The point of this litany, Barlow says, is to emphasize that "this is not a drought in the sense of the old days, when shortages meant an interruption—and then resumption—of normal rainfall. This is the end of water."

Barlow cites the most recent academic studies and fresh water experts to back up her conclusions.[3] She explains that "we were taught the hydrological cycle can't be disrupted, but that's only true if water is returned to its *source*. We're removing water from its normal sheds and storage areas, and it ends up in the ocean or dumped in such a way that the rain cycle doesn't return it to its original area. Today we're increasingly moving it from where it needs to be to where we want it." In the process of taking water from one area to feed the ever-growing thirst of our industries and cities in another, we are increasingly disrupting and destroying the hydrological cycle. Think of the 14 million people of Mumbai and Shanghai, the 13 million of Tokyo, the 20 million of New York, then add 10 for Beijing, 8 for Kinshasa, and 12 for Delhi. The farmlands, forests, or marine wetlands on which these cities and their sprawling suburbs and industrial parks were built have been paved over with cement and asphalt and sealed by buildings. All the rainwater that falls in their precincts drains immediately into nearby oceans or big rivers instead of being absorbed by land, where it could have returned to the cycle as usable "green" water. Now, tainted with city pollutants, it becomes saltwater and is lost to humans, and everyone suffers. New Delhi, for example, is experiencing daily riots as the water tankers that are the only supply for residents in the poor south and western parts of the city either don't arrive or don't have enough to go around.[4]

Allerd Stikker of the Amsterdam-based Ecological Management Foundation explains how things came to this pass very simply. "The issue today... is that while the only renewable source of fresh water is continental rainfall, the world population keeps increasing by roughly 85 million per year. Therefore the availability of fresh water per head is decreasing rapidly."[5] Agriculture uses close to 70 percent and industry 25 percent; less than 10 percent goes to households and municipalities. The amounts in these three categories have been doubling every twenty years, which is more than twice the increase in the population; that means wastage and profligacy are at work. The High Plains Ogallala Aquifer, for example, is being depleted eight times faster than nature can replenish it. This water feeds the luxury crops of fruit and flowers being grown in California semideserts; it fills swimming pools all across the American Southwest; it runs lawn sprinklers so that grass grows in the scorching desert; and it circulates in air conditioners, some of which cool outdoor patios so that their owners never have to contend with the natural climate. The net result is that the water table under the San Joaquin Valley, the epicenter of California agriculture, has dropped about thirty feet in many places over the last fifty years—this is water that has been permanently lost.[6]

The deep-well drilling systems we greeted with such joy sixty years ago are now being called "water mining." Not only do they deplete the red water that used to underlie whole ecosystems and keep them stable; they cause salts to invade freshwater aquifers.

When it comes to what are called "fossil aquifers," ancient underground water storage areas, relentless water mining with drilled wells can cause the caverns and stone hollows holding the water to collapse, permanently reducing the Earth's water storage capacity for the future. What no one realized back when all the drilling started, as a report from a highly respected research NGO, the Third World Network, explains, is the fact that although some groundwater aquifers recharge quickly, "the *average* recycling time for groundwater is estimated at *1,400 years*, as opposed to only 20 days for river water."[7]

There has also been an enormous increase in the amount of "virtual water exporting" since the theory of economic globalization overtook the world about twenty years ago. Israeli economists first used the term *virtual* or *embedded* water in the early 1990s, when they realized that growing oranges or avocados in the desert to trade for other goods in the global economy meant that they were in effect exporting scarce Israeli water resources with each crate of fruit—and at bargain-basement prices. Irrigation, the mainstay of mass farming in sunbelt areas, is a stunningly inefficient use of scarce water. More than half is lost to seepage or evaporation in flood irrigation. The process also leaches out salts and minerals that will eventually destroy the soil, which is why so much of the once-fertile crescent in Iraq and Iran is now useless, salinized desert. When grown by irrigation in a hot country, crops like lettuce, fruit, flowers, or cotton remove astronomical amounts of fresh water from each watershed. It will be consumed and excreted as waste thousands of miles away, so very little of it will return to the farming area as rain. A small bag of salad lettuce requires 80 gallons of water to produce. A single pound of cotton can take up to 3,600 gallons of water to grow.

As we all know, rarity increases price in our money-based economy, and within the last ten years the proof that water is becoming threatened is that it has turned into a very desirable commodity. Coca-Cola, Pepsi, and water giants like Vivendi, Bechtel, and Suez are doing their best to corner the market on this necessity. They know they'll be able to charge almost anything for access to pure water. That's why a major focus of the Council of Canadians, Barlow's new NGO the Blue Planet Project, the International Union for Conservation of Nature, and the UN Human Rights Council, among many others, has been to fight the privatization of water. Reasons to fear for-profit, corporate control of the world water supply have grown and have been supported by experience in many arenas over the past decade. In the huge majority of cases, the big water companies' promises of greater access for poor people, lower rates, and better service have not materialized.

With water facing such alarming prospects, it's very good news to have Maude Barlow speaking up for all the living things that depend on it. Her appointment to the UN in 2008–09 demonstrates a dramatic change in what the powerful are willing to learn about the state of the world's water. And in some surprising parts of the world, the first glimmers of intelligent local and national policies are already beginning to appear.

THE RIVER MAKER

This is the story of a man who did not go to poverty-stricken villages in Rajasthan with the intent of "developing" the "under-developed"... He went in all humility to find out if he could be of service... he simply put himself at their disposal.

Profile of Rajendra Singh in *India Together*[8]

Back in the 1970s when the late Israeli scientist Michael Evenari
was studying ancient methods of rainwater storage in the Negev
desert, he discovered something very significant. Evenari's landmark studies proved, to everyone's surprise, that a small watershed
will produce more water per acre than a large one. A watershed of
2.5 acres provided as much as 10,000 gallons of water per acre per
year, whereas an 850-acre watershed yielded only 2,500 gallons,
about a quarter as much. Evenari's findings help explain why small
water projects that mimic natural systems and take advantage of
local customs and conditions are more likely to succeed.

In central and northern India, for example, rainfall is sufficient,
but it comes all at once in the six or eight weeks of the monsoon
season. This feature makes it a typical "brittle environment," the
classification developed by Allan Savory that means special care
has to be taken so that it can maintain its fertility. That care wasn't
taken. Areas like Uttar Pradesh were heavily forested within living
memory; in the 1940s people could still hunt tigers in forests that
are now mere heaps of gravel, with sad cows and goats nuzzling
garbage or dead grass on the barren slopes. Like so many parts of
southeast Asia, one generation of greedy deforestation by the British and cash-strapped, newly independent India, combined with the
area's naturally poor, rocky soils, means that this land can no longer
hang on to its rain. The life-giving downpours crash into temporary
torrential rivers that tear past the thirsty plants and inhabitants,
carrying the water off to the sea. After a month or two, every ves-
tige of water has vanished, and the wide scars of the bleached, dry
riverbeds are the only sign of its passing.

Adding considerably to the problem is the quick fix of modern
water mining. From the 1960s to the 1980s, well-meaning government agencies and foreign NGOs ran all over India sinking deep

wells, to solve the immediate problems caused by deforestation and mismanagement of the rain resource. The population quickly became dependent on these wells for drinking, animal needs, and irrigation. Today the levels of groundwater—that's the red, do-not-touch water—have dropped precipitously. There are places on the Indo-Gangetic Plain that have supported a large population of agricultural peoples for at least seven thousand years where people now have to drill 900 feet to find any water. The Punjab, India's rice basket, where 1.5 percent of the country's land mass produces 50 percent of its grain, is so punctured with wells it looks "like Swiss cheese." "We have depleted the ground water to such an extent that it is devastating the country," says Gurdev Hira, a soil and water expert from Punjab Agriculture University.[9]

Some of the riverbeds in central and northern India have been dead for a hundred years. Monsoons and poor soils have always been a feature of the area, but prior to British colonization the rural areas of India had benefited from complex systems of water harvesting and storage that varied according to topology, need, and local customs. One of a maharaja's typical responsibilities was to provide water by building and maintaining the largest structures needed for rain- or green-water entrapment, including small dams, water tanks, artificial lakes, and other expensive installations. Local people contributed their labor and also kept up smaller irrigation and water confinement systems on their own through a practice called *goam*, voluntary labor.

In the past, besides building artificial tanks and cisterns, people used what they call a *bund*, or a "check dam," so called because it briefly checks the flow of water in the rainy season. These are very similar to the kind of dams beavers make, and in that sense they mimic nature. A free-flowing stream is blocked with earth and rocks, sometimes with concrete, to create a small lake above the obstruction so that the amount of water that flows below the dam can be controlled. Like beaver lakes, bund impoundments not only store water for many months more than it would remain without them, especially in the absence of forests, but also allow it to percolate down through the soil into the groundwater and deeper aquifers, thus replenishing wells for some distance in every direction.

Tarun Bharat Sangh, or TBS, is an NGO operating in Rajasthan, one of the poorest and most desertified areas of India. It's the vision of one man, Rajendra Singh, who began his search for "something to *do*" about poverty when he was a young health and social service worker in the 1980s. His first experiences of working with nomadic tribes in this northwestern state taught him to prioritize entire ecosystems—healthy wildlife and forests—*equally* with people and crops.

This means that TBS bases much of its understanding of ecosystem needs, like so many other effective environmental organizations, on what we've been calling traditional management, aboriginal science, or adaptive management. Singh uses a form of adaptive management to reach his extremely lofty goals, another sign of a sustainable system. TBS's website uses only eighteen words to explain the organization's two-step model for restoring pretty much everything, from crop yields to cultural values and personal self-respect: "First, revive vegetation on barren hill slopes. Second, build small water catchments in the valleys and the plains." But how do you get people to stop pasturing their cows and goats on any hill when they are hungry pastoralists whose only wealth is their starving animals? Singh's NGO discovered that "No amount of money, government action, or legislation can deliver results... only people's fullest cooperation can achieve these ends." How do you get that cooperation? TBS uses the common tribal and traditional peoples' method of *total* consensus, which requires, as the title of one section of its website says, "Talk, talk, talk."

A major TBS goal is usually to construct a *johad*, an ancient water storage method similar to a bund or check dam, which Singh learned about from village elders. These earthen water tanks are like enclosed beaver ponds. Inlets help guide the monsoon's overabundance of water into the johad and let it gradually percolate down to recharge wells and rivers. Singh says that "the design, location, and construction of each water-harvesting structure is discussed endlessly [within each village] until a true consensus is reached." Consensus is required because the soil type and location must be exactly right, meaning some villagers may have to give up their fields to make room for it and then be compensated, and especially because every single inhabitant has to agree "to contribute either money or labor... to the construction of a johad, a check dam, or a weir."

Singh admits achieving consensus is the hardest part and that one village took five years to hammer out their agreement. "To the modern mind, that may seem too long for a piece of civil works that then took only about six months to build," but once such community ponds are built, "they become 'everyone's' " and are guarded and maintained religiously. "Issues of use and sharing have been settled before construction began, rather than later. Such works are forever, and the five years of deliberation recedes into insignificance."

Villagers at TBS meetings don't just talk about goats and hills or water dams. They have "endless discussions on every conceivable issue," including problems of drunkenness, spousal neglect, outmigration, loss of bird life, animal predation, children's health, use of medicinal herbs—you name it. What TBS has discovered is that when you restore water, which restores agriculture, animal husbandry, and economic prosperity, you are also restoring community and culture. And that "when small communities like these succeed, the government itself wakes up and development becomes what it should be: ground-up instead of top-down."[10]

Here's one example: In the early 1990s drought had reduced the villagers of the Rajasthan village of Nimbi, near the city of Jaipur, to poaching game from a nearby forest or working in the city; the rest of the time the men drank, in despair over not being able to sell their land or get it to produce. Family and cultural life were at a nadir. The desperate village elders approached TBS, which helped them set up a holistic plan that began with "talk, talk, talk." That plan first of all demanded a voluntary ban on drinking and poaching. With a small financial start-up from the NGO, the now sober villagers raised internal money, built several check dams, and renovated their old, silted-up johads; they planted trees and started an organic dairy. Within just a few years wages increased by 300 percent, land prices shot up, they had a moisture-conserving forest, and the farmers could produce up to three crops a year, using completely organic methods. Nimbi has become a source of produce and dairy products for the nearby city, and no one has to poach for a living or migrate anymore. Sounds quite amazing, but that's only half the story. Such villages usually find more enemies to contend with than mere droughts and drunkenness, and TBS and the villagers it works with have established the following rule: "Keep the government at bay, with defiance if need be."

Unlike most NGOs, which tend to mimic corporations in seeking to get bigger, which almost inevitably leads to partnerships with governments and industries, TBS has been careful to stay small, bottom-up, and decentralized. The organization has now built over four thousand johads and helped build another five thousand in over a thousand villages in a 2,500-square-mile area. Most spectacularly of all, "Five seasonal rivers—the Ruparel, Arvari, Sarsa, Bhagani, and Jahajwali—in the northeastern Rajasthan area that had nearly dried up, *have now become perennial*."

In the case of the largest river, the Arvari, "which had altogether disappeared during this century... the miracle happened quite unexpectedly, after eleven years of water harvesting and growing new tree cover." The grateful villagers of nearby Hamirpur forbade catching the fish that had also miraculously reappeared and even devised special rituals to encourage feeding and protecting the new aquatic life. However, it shouldn't surprise anyone who deals with regional governments that in 1997, "as soon as these rivers had started producing again, the state government stepped in and... the Fisheries Department of the Rajasthan government began to issue fishing contracts to commercial interests from outside the area." The people were outraged and blockaded access to the river. Their ally, TBS, was threatened with dire legal consequences if it supported them. "It responded by asking the government to show proof that it had ever before awarded fishing contracts." Thanks to the people's connection with a savvy and incorruptible NGO, the issue got media coverage and "the government had to withdraw and cancel the fishing contracts." But it didn't end there. Having nearly lost their precious river, the people of Hamirpur "decided to institutionalize their rights to protect and safeguard the river, not only from government encroachment but also from individual greed."

Already a few of the locals who had been forced to move away had returned to the area and were siphoning off water to irrigate their fields, a practice that was rapidly reducing the river's flow. The villagers knew it wasn't sustainable. Today, ten years later, there is an Arvari *sansad* (parliament) with 142 elected representatives from every village in the area, which has encoded comprehensive and detailed rules for water use. Watershed users in Canada or anywhere else in the world should gape in envy at what has been enacted by a group of supposedly "illiterate peasants," in one of the poorest parts of the planet.

First of all, water-intensive crops, like sugarcane and rice, are forbidden. "Crop choice and pattern [are] determined by the total available water resources in the area, rather than on individual [or market] whims." After the spring festival of Holi, no one can directly take water from the river; this restriction allows it to survive the hot summer months. The rest of the year, anyone wanting to use water for irrigation must pay a small but significant fee to the village cooperatives. "While there are no restrictions on people selling their land to other farmers, no one may sell his or her land to outsiders for industrial use because that will mean loss of power for local village institutions to self-regulate."

There are also strictly observed codes of conduct to prevent grazing practices that might adversely affect any other plants or animals. No species is seen as too small or too peripheral to support a healthy watershed. The Arvari sansad acknowledges the interdependence of all parts of a natural system with a universal and inflexible rule: "No one is allowed to hunt wild animals in this area." The hunting ban has proven so successful that "many of the wild animals from the government-controlled Sariska Wildlife Sanctuary" nearby have migrated to the "peoples' protected forest," thereby making their ecosystem more viable in every way, restoring small biota along with the big animals.

TBS stands by its villages and everything they try to do to better their surrounding ecosystems, no matter what. Rajendra Singh has been arrested and beaten several times by local government agents. In 2002 he suffered skull fractures when an Uttar Pradesh district functionary attacked him with a club for "inciting the public to disobedience," by telling a public meeting, "If the government does not help, one should learn to help oneself." When TBS began wantonly planting trees and building johads on government land to restore water to the Sariska Tiger Reserve in Rajasthan, the state charged the group "for infringing the Wildlife Protection and Forest Conservation Acts" no fewer than 370 times. TBS kept at it, and the reserve, down to five tigers in 1988, now has twenty-seven. Singh and TBS also fought the illegal marble mines within the reserve that continue to drain water from the area and despoil the forest. They went to the Supreme Court of India to enforce a ban, during which time Singh survived three attempts on his life by mine owners. But despite gaining closures, TBS didn't forget the now unemployed miners, helping them reestablish themselves "through water and forest conservation activities and rural development, like animal husbandry, agriculture, [and other] employment generation activities." As recently as 2006 the district administration demolished TBS buildings that had been erected near the Arvari river ten years before, but TBS went to the media and then the courts for compensation and the right to rebuild.

Today, despite TBS's fierce political independence, decentralized structure, and outsider status, Rajendra Singh is being hailed in the media as one of India's "leading environmentalists" and was recently named a member of the country's prestigious Artificial Recharge of Ground Water Advisory Council, part of the Central Ground Water Board.[11] No fewer than five northern state governments have come, hat in hand, to this grassroots-based NGO to humbly ask for advice for irrigation, forest, and watershed programs.

The Alwar region, where so many rivers have been restored, sees tens of thousands of official visitors every year, who come to see the miracle of river rebirth. The villagers themselves have received numerous national awards, while Singh was honored with the Asia-wide Ramon Magsaysay Award for community leadership in 2001. Best of all, these ideas have caught on across India. Tens of thousands of similar water entrapment structures are being built in the same small, self-financed way, each designed to accommodate specific local needs.

Yet, on a wider scale, the central government, more often than not, works to discourage and thwart these proven methods by constructing megadams like the Narmada, which displace millions of people and permanently destroy entire ecosystems. That's because India's government, like so many others, sees globalization economics and large-scale technical fixes as their main source of future prosperity and has been slow to recognize the long-term superiority of more humble approaches. Another factor is that no one, especially no one from outside the area, stands to make money on check dams and other old-fashioned water-impoundment techniques. They're not even on the radar of the big funders like corporations and the World Bank, which have billions of dollars to allocate every year and naturally favor capital-intensive projects.

India has more working examples of water crisis solutions already in place than North America is even thinking about, probably because so many of its people are in a position where they have to develop them—or die of thirst. But desperation isn't all that's needed. A culture that understands the value of ecosystems and enjoys at least a modicum of democracy are key requirements. China has also used up or contaminated almost all the water it has in its quest for industrial primacy. Its response so far has been short on humility or flexibility, no doubt because of its political structure. Rather than small-scale rain harvesting, it's even more likely than India to favor top-down attacks on the ecosystem, megaprojects that will destroy the hydrological cycle more thoroughly. China's current plan involves robbing what's left of Tibet's melting glaciers and piping that water thousands of miles across China to feed the eastern and southern factories that will pollute it before the whole mess runs into the sea. It's not hard to see which approach—China's or that of Rajendra Singh—has the potential to save the world's hydrological cycle over the long term.

NOT WASTING WHAT YOU'VE GOT

If you're short of water, the choices are conservation, technological invention, or the politics of violence.

Marq de Villiers[12]

Bringing water back from the dead, as they're doing in India, is one effective way to make sure there's enough to keep the ecosystems and people going, but it demands actual rain. In many parts of the world that were dry and barren long before central India became desertified, conserving what little green water falls from the sky has become critical. Some time ago Joyce Starr of Washington's Global Water Summit Initiative pointed out that "Nations like Israel... are swiftly sliding into that zone where they are using all the water resources available to them. They have only a decade or so left before their agriculture and ultimately their food security is threatened."[13]

Jordan is already there. Jordanians are allowing good agricultural land to lie fallow because they don't have the water resources to irrigate it. By 1995 all known sources of water across the country were already being tapped, and the demand for more is growing rapidly, owing to an influx of young refugees from Palestine and a general increase in the standard of living. Other countries in the Middle East have almost no water. For example, the United States has an estimated freshwater potential of 2.6 million gallons per citizen, per year. Iraq has half that, Turkey only 1 million; but even Egypt's potential of 290,000 seems luxurious next to Israel and Jordan's 120,000 and 69,000 gallons per person, respectively. But because of their limited supply, both of these countries have become extremely innovative in their use of water.[14]

Like the people of Uttar Pradesh and Rajasthan, the ancestors of the Jordanian people knew how to harvest rainwater. Archaeologists have discovered that the ancient Nabataeans, an Arabian tribe that built beautiful cities and temples in this region 1,600 years ago, used many of the varieties of water entrapment methods that have been observed in India, as well as huge cisterns and stone water channels hewn directly into the cliffs. The Nabataeans even carved inscriptions alluding to the need to store and conserve water. One Nabataean irrigation method was to build low walls around small plots of land to trap runoff and then to plant crops in the boggy soil; those crops grew as the water seeped into the ground.

When the modern era introduced deep-well drilling that was able to tap into red-water aquifers, the region's intricate system of cisterns, reservoirs, and collection channels for catching rainwater collapsed, just as India's did. Now that aquifer and groundwater levels are dropping fast, several Nabataean methods are being revived. The low-wall method is again being used in Israel, at the instigation of the Ben-Gurion University of the Negev; Jordanians have also realized the folly of tapping groundwater and fossil aquifers. Today they're continuing to develop modern versions of rainwater entrapment and efficient use of irrigation that will help them recharge aquifers and also not contaminate what little rain they get.[15]

Irrigation has been used for thousands of years to get water from lakes, rivers, and wells out to the crops. As Marq de Villiers says in his book *Water*, "Where drainage is good and the soils not naturally alkaline, carefully managed irrigation can persist for centuries without harm. In areas where irrigation is used as a water supplement rather than the whole diet, little harm is likely to befall the land."[16] But irrigation has a real catch to it when people try to grow crops in a desert, where the soils are poor and drainage is inadequate. Over-irrigation and poor drainage, common in places like Uttar Pradesh, the American West, and the Middle East, where rain comes all at once or not at all, carry away runoff filled with agricultural chemicals and naturally occurring minerals and salts. The reason oceans are salty is that all water leaches out the minerals in soil and carries them to the sea. Synthetic chemicals in substances like asphalt, spilled oil, and herbicides are swept up and carried along as well.

The salts and chemicals end up precipitating on the surface of soils when the water used for irrigation evaporates in the desert sun. There they kill soil organisms and render the soil sterile. Moreover, when fields are poorly drained, the water table rises up and keeps the soil too waterlogged for plants to grow properly. Added to these problems is the relatively recent addition of petrochemical fertilizers, which destroy all organic matter, and the use of pesticides, which also destroy organic life in the soil and threaten human health as well. Together, irrigation and industrial farming are destroying fertile agricultural land at a very alarming rate. Millions of acres of land across the world have been damaged by salinization and contamination, and millions more are in danger unless we figure out a better way to move water around. Hybrids and genetically engineered seeds require a lot more water than traditional varieties, and despite claims that biotech will create drought-tolerant crops, none are on the horizon. We can't continue to plant water-guzzling crops in the belief that technology can bring water to desertified areas without any long-term costs.

We need to start by managing irrigation better. About fifty years have already passed since engineers at the Technion-Israel Institute of Technology in Haifa came up with the method called "drip-irrigation." It uses perforated hoses to deliver much smaller amounts of water directly to the root-zones of the plants. In this way both runoff and waterlogging are prevented; 95 percent of the water goes to nourishing the crops, as compared with only 20 percent or less with conventional irrigation; as for spray irrigation, as much as 92 percent of it can be lost to evaporation! Even though drip irrigation also increases crop yields by between 20 and 50 percent, only about 2 percent of the world's irrigated fields are using it, a figure that hasn't increased much in the past few years. The subsidies that today fund ecologically dangerous big dams, poisonous agrichemicals, and hefty royalties for dubious genetic technologies could be redirected to help farmers purchase or lease equipment for drip irrigation.

Not wasting water doesn't just apply to irrigation, of course. Water use involves entire communities and requires well-thought-out policies and legislation. Every country, every municipality worldwide has had to develop rules concerning water access and allocation. Most places in the world, like Mexico and India, guarantee free access to shorelines, but that's not true in the U.S. or in Canada.[17] Every major body of freshwater today is the subject of allocation agreements among the surrounding populations concerning access and the amount that people can remove for private or municipal use or for irrigation.

Military and political analysts have long predicted that water shortages will lead to terrible international and internal conflicts, and political columnist and author Gwynne Dyer's most recent book,

Climate Wars, is one of many analyses preoccupied with the idea of conflict over water.[18] Dyer imagines some future war scenarios, including a desperate war between Pakistan and India over the flow of the Indus River, already underallocated to Pakistan. But despite the talk about future water wars, so far, places like Israel, Palestine, and Jordan have managed their shared sources in a relatively civilized manner, even in times of armed conflict. For example, during the 2009 war in Gaza between the Israelis and Hamas, the Friends of the Earth water-sharing committee continued to meet. There really is inspiration to be found in the way some people, even longtime enemies, recognize each other's need for water.

Legal mechanisms for peaceful sharing are already in place in most of the world; people just have to figure out how best to use them. Legislation and treaties abound with terms like "reasonable use" and "prior occupancy," which work well in times of plenty but need to be adjusted in order to protect stressed river systems or failing aquifers. Legislators are evolving older ideas like "beneficial use" into new ones like "nonmarket use," which means using water supplies to try to cover not just human needs and desires but also the vital and basic needs of the surrounding ecosystem. However their laws are codified to cope with the new reality of water scarcity, the world's most water-pressed people are going to have to learn to make the wisest possible use of whatever water they do have, as quickly as possible.

They've made a start. In India and Jordan rainwater harvesting is the law for all new construction in several regions, and nearly all the big cities, especially on larger buildings. New buildings that don't have rainwater-harvesting technologies are outlawed in Singapore, a place that's so encased in cement and asphalt that 50 percent of its water is already imported, and so, such policies are absolutely necessary. Even China has begun some rainwater harvesting, especially in the poorest areas where there is severe lack of access to other sources. Traditional clay-lined *shuijiao* (water cellars or cisterns) are being upgraded and new ones constructed. Sri Lanka is reviving its traditional water-harvesting technologies, using cheap materials like tin roofs, bamboo gutters, and banana-stem pipes to guide the rain into cisterns, tarps, or rock depressions.

These methods are largely for collecting drinking, washing, and toilet water; when it comes to agriculture, there are dozens of ways that farmers can trap water so that it lasts a little bit longer. Contour ridges placed cleverly along slopes can trap water the same way a lip on a staircase would. In East and West Africa, small pits are dug across fields to trap rainwater so that it can gradually feed the crops instead of causing erosion. In Tunisia *meskats* in fig and olive orchards use catchment areas and conducting channels to direct the water around the trees. In Pakistan hillside-runoff systems called *sylaba* direct the water cascading off a rain-drenched hillside to fields surrounded by levees.[19]

Why is every country not dotted with such simple devices and structures? Dry areas in North America would all respond beautifully to the same kinds of traditional rainwater conservation efforts described here, but they are rarely practiced. One reason is that they don't make profits for anyone, they don't advance anyone's political career, and they require human rather than machine labor. That's why societies and economies built on romantic notions of never-ending wealth don't like to use them. But if we carefully consider physical realities in our management policies, we would see that these systems are perfect. Cheap to build and simple to maintain, they work *with* the hydrological cycle. They have been proven to be able to rebuild entire river systems and watersheds. Rather than apply them, rich countries, or the elites within poor ones, usually use their wealth to steal water from other places, assuming it will somehow find its way back. If that's how society continues to manage water, then endless conflict, water refugees, and full-scale water wars *are* inevitable.

Maude Barlow's primary assessment of how the planet's hydrological cycle works is that "water is where it is for a reason; it should not be moved!" It may seem unfair to humans that the northern part of North America is as much lake, river, and bog as it is land, while a thin stream in Utah or Arizona is isolated by hundreds of miles of parched desert. That imbalance looks like an opportunity for human intervention, bringing the waters of the Colorado River or even the Great Lakes west or south to feed the barren valleys of California and the dry Midwest. But the hydrological cycle, like every life cycle on this planet, is very complicated. We're just now learning that it depends on *where* the lakes and rivers are located in the first place: in warm areas or in cold ones, on rock or soils, inland or by the sea. What happens when engineers change the direction of an entire watershed so that it drains in the opposite direction, as they have done with so many of our big dam complexes? What happens when we suck up so much of a big river like the Colorado or the Indus that its delta dries up and it can't make it to the sea? Scientists have no idea why there is so much water underground and what it's actually doing down there. We are starting to understand, however, that there is almost nothing in nature that doesn't perform a vital function. Could it be regulating temperature or lifeforms we don't know about, like all the extremophile bacteria that have been so recently discovered? When we drill down and pump deep-stored red water up to the surface, what exactly are we doing to the whole system?

Communities have learned the hard way what happens when water is moved away from where it naturally occurs.

Biodiversity—and all the services those life-forms perform—vanishes from drying, salinized estuaries. Local weather systems change radically when a lake either appears behind a dam or disappears after it's drained for irrigation. When people dam huge rivers that flood peat bogs to create power, as Canadians continue to do in northern Quebec, we release mercury into the food chain where it poisons every living creature. Slight changes in the temperature of ocean currents off the coast of one country, Peru, can create huge floods and droughts that rage from North America to Africa, as they do so often with El Niño events. So what do all these perturbations on local levels do to the hydrological cycle? We're only just beginning to find out, and it looks like the gut instincts of people like Maude Barlow are right. Fresh water is where it needs to be on this planet to feed the wheel of life, maintain the atmosphere, regulate the weather, replenish the oceans, and nurture the forests. Move it around, spread it out, cut it off, suck it dry, and this sublime machine just might break down. Many people know this, deep down. That's why even some water-rich countries are taking steps to make sure they stay that way.

GREEN VERSUS GRAY

Green infrastructure understands that development and growth are inevitable... [but that] its location, form, and intensity can be directed and managed... proper green infrastructure systems lessen the need for expensive gray infrastructure installations.

North Carolina urban development website[20]

It's instructive that in Germany, where the rainfall is high, the vegetation is lush, and groundwater is plentiful, the same kinds of water conservation precautions common in the deserts of Jordan or Rajasthan are being introduced. In most of Europe, for example, toilets are routinely manufactured with a choice of two flushes—a small amount for urine, and a larger one for, as the Germans put it, "the bigger business." Low-flow showerheads are the norm, and easy-to-read water and electric meters are placed prominently in homes and apartments so that people can keep track of what they're using. "Gray" water is often recycled automatically by village or city systems, and there are regular inspections of urban water delivery systems for leaking pipes, which cause an amazing amount of waste, contamination, and illness.

Even in places like the Pacific Northwest of the United States, where water would seem to be over- rather than underabundant, cities have begun to implement strategies to make sure it's not wasted or polluted. The problem in every city, even a small one like Portland, Oregon, is that storm runoff picks up contaminants from tar roofs, car parks, sidewalks, and industries and then sweeps the polluted water into municipal sewage systems that overflow and contaminate the rivers. Too much water, in other words, can also cause contamination and waste. The "grey infrastructure," our parking lots and buildings, don't allow it to soak into the soil and replenish the groundwater any more than hardpan or denuded deserts do. In fact, as a recent *Frontline* documentary, "Poisoned Waters," put it, "Stormwater runoff is the worst threat to our water... Seattle alone carries an Exxon-Valdez level of petrochemicals into Puget Sound *every two years*."[21]

Erik Sten, who was sustainability commissioner in the city of Portland, along with Dan Saltzman, who serves in that position now, instituted a stick-and-carrot approach to conserving rainwater and maintaining stormwater purity back in the early 2000s. Saltzman explained, "Property owners who want to build green roofs to absorb and purify the rain are given subsidies and tax breaks." Homes are also offered grants to make sure that roof run-off is channeled into the lawn or shrubbery before it hits the sewers. "Or we do the whole thing ourselves, and places that do nothing can be subject to fines." The green roof concept has grown rapidly in both Europe and North America. Greenery looks great on the flat roofs of big apartment or office blocks and sometimes serves as a little park for the tenants or a place to grow some vegetables. Chicago now has 2.5 million square feet of green roofs on two hundred buildings, more than any other U.S. city, and these roofs save each property owner an estimated 20 to 30 percent of their heating and cooling costs every year.[22]

Using plants to absorb, cleanse, and conserve water is even more effective when the greenery girdles urban areas. These widely used greenbelts exist for two reasons: to preserve some of the watersheds and biological services that the city has destroyed and also to protect the areas close to cities from urban sprawl and further environmental problems. Modern greenbelts vary widely, from a few connected parks in cities like Stockholm, to Frankfurt's ambitiously thick swath of organic farms that supply the city with food as well as water purity and wildlife habitat. The states of Oregon, Washington, and Tennessee have legislated "urban growth boundaries" for their largest cities, in order to preserve watersheds and green spaces. City governments have also mandated growth boundaries in Toronto, Ottawa, and Vancouver in Canada, and in Minneapolis, Virginia Beach, Miami, and Anchorage in the U.S. There are large greenbelts around Adelaide, Australia; Sao Paulo, Brazil; and Seoul, Korea. There are also greenbelts that encompass entire regions, like the 1,980-square-mile greenbelt around London.

Today the greenbelt concept has been joined by the term "Green infrastructure," which includes "urban forests." These are now seen as a way to help "get the rain out of the drain" and keep raw sewage out of our lakes and rivers.[23] Cities can't function without patches of woods and tree-lined streets, plant-filled yards and empty lots, parkland and remnants of farmland. This greenery is needed to balance the "gray infrastructure"—the asphalt, cement, and buildings that pollute and redirect water, destroy habitat even for soil biota, absorb too much heat, or otherwise reduce ecosystem services.

Researchers studying urban forests throughout the world to determine their "structure, function, and values" have been impressed by the pollution sequestration and air and water quality services provided by even small, straggly woodlots and yard or curbside trees. The town of Oakville, Ontario, discovered that its yard trees saved residents $812,000 annually in reduced energy bills and absorbed twice the amount of sulfur dioxide, as well as all the particulate matter, generated by the city every year![24]

One province in Canada has unconsciously led the way in terms of how a really big green zone can help a city retain farmland, habitat, air quality, and its water supply. Quebec's remarkable "green zone" legislation is a legacy of René Lévesque's attempt to ready the province for independence in the mid-1970s. It froze most of the arable farmland in Quebec south, east, and west of Montreal, creating one of the largest greenbelts around any city in the world, about 5 million acres. Of course, its efficacy as a water regulator is regularly encroached upon by developers looking for loopholes and increasingly industrialized farming and dumping practices. But it shows what can be done in terms of legislation and especially what rural and suburban property owners will accept. Few Canadians or even urban Quebeckers realize that virtually *no development* is allowed in Quebec's enormous green zone unless it directly services the needs of agriculture. That means no selling off even one little building lot, no putting up a cottage for the relatives or even a spare garage, without a lot of provincial oversight. No subdivisions, no strip malls, no trailer courts outside of municipal boundaries. This law was passed in 1978, more than thirty years ago, and these days the strongest support to retain it comes from the once-resistant farm and rural landowners. They now fight any top-down attempts by the provincial government to weaken it—a sure sign that it's recognized as sustainable for the land by the people closest to it.

SAVING MONEY THE HARD WAY

There may be scope for reducing net outlays to veterans by removing the tax exemption for benefits, and for slowing the growth of transfers, to Indians and Inuit... federal funding for research [and oversight] is extensive... there would seem to be scope for rationalizing these services *with a view to increasing the private sector's responsibility for such activity.*

IMF Structural Adjustment Programme for Canada, 1995[25]

Canadians still remember that in a prosperous, rural part of their country, seven people died and more than two thousand others became seriously ill, simply because their drinking water was contaminated. What affected half the population of Walkerton, Ontario, wasn't a freak chemical or pesticide spill. Instead it was a combination of unsustainable industrial farming practices and government cutbacks to water-testing services. When feces-contaminated runoff from a large cattle feedlot entered one of the town's wells during an excessively rainy spring, inadequate testing procedures failed to identify the problem: deadly *E. coli* in Walkerton's water.

In the prosperous First World, we well understand the dangers of bad water, and over the last century we've instituted expensive chains of regulations and inspections, as well as chemical purifiers, to make sure people can use town and city water with confidence. Food and water have always been sources of disease, but by the 1950s most parts of the First World had contamination by feces and bacteria well under control, usually by means of government regulations backed up by state-funded, regular inspections. However, an integral part of modern economic theory is the strange idea that agencies providing vital public services like education, health, and police protection should be economically self-sustaining. Especially since the Reagan/Mulroney era, it has become an accepted idea that services like health care, education, or food- and water-testing are "too expensive" when run as nonprofit public services by state or municipal governments. Although there is *no empirical data* to support this claim, private, profit-making companies have been credited with being "leaner" and more efficient and were assumed to do a better job at delivering almost any service.

Such public-private partnerships were mostly forced upon governments by major financial institutions, including the International Monetary Fund (IMF), as part of globalization's "structural adjustment programmes," when governments got into debt in the 1980s. This reordering of national priorities to pay off international debts has destroyed social services in hundreds of countries, including Canada, where such services, including health care, are increasingly denied funding, closed, or privatized. Most of us don't know all this because IMF reports to indebted governments are secret and can only be obtained through Access to Information laws. Canada and the U.S. are most definitely debtor nations, even more so since the economic downturn of 2008, and as such are told every year by the IMF how to service our debts—even as our tax dollars help prop up this highly unpopular institution.[26]

In order to comply with the demands of such lenders and in the belief privatization would maximize efficiency, the Ontario government, along with many others across Canada, began taking itself out of the water- and food-testing business. As early as 1997 the government started to hire for-profit private companies—in this case, A&L Laboratories in far-away Arkansas. The Ontario government distanced itself from water safety even further by *not* requiring the company to notify health or environment officials of any irregularities; they only had to notify the municipal waterworks manager, who in this case was largely untrained.[27]

Under the public system in place just a few years previously, all Ontario water was tested locally and at regular intervals by a fleet of trained government inspectors who had to report any contamination to both local managers and the provincial health ministry as soon as they detected it. That system worked well, as Maude Barlow, an Ottawa resident, points out. "We really did have a good system in Ontario. We had watershed management, we had strong testing in the regions. There probably were areas that could have been improved, but instead of getting stronger, what we used to have has been steadily deregulated, piece by piece. Testing was cut back and regulations on corporate farms and industries have been smashed, so as not to 'interfere with their economic growth.' But there's absolutely no reason why a government system that is adequately funded and politically supported would not have continued to give us pure water."[28]

Infection from the organisms that invaded Walkerton's water involved more than simply getting diarrhea and then getting over it. Two of the small children affected have developed a severe platelet disorder similar to hemophilia and may never lead normal lives. Many other residents were traumatized by the painful, invasive treatments required; and of course, seven people died.[29] But Walkerton's deaths could have been easily prevented, with measures that are almost ludicrously simple.

FOULING YOUR NEST

The sheer number of dangers associated with treating sludge as if it were a fertilizer is so great, so various and so serious that it would be the life work of thousands of professionals to divide up and respond to the categories of problems that will arise from this practice.

Scientific report on sludge, Cornell Waste Management Institute[30]

Of course organic contaminants like *E. coli* are only part of the problem. The petrochemicals we burn, the heavy metals we mine, the plastics and other nonbiodegradables we throw into landfills, even the medications we take to stay healthy, have all contaminated our water, putting it into the "black water" category. Right now most provinces in Canada and many states in the U.S. are finding themselves saddled with a never-ending river of toxic sludge: human feces and urine, mixed with industrial chemicals, prescription drugs, heavy metals, and any pathogens the humans may have carried. These municipal sludges are turning out to be the biofuels of water: "green" intentions that have backfired because they weren't well thought-out.

Back in the 1970s, jurisdictions across the world realized they could no longer dump raw sewage and industrial effluent in their rivers, lakes, and oceans and still have safe water to drink. The water treatment plant was invented, and engineers quickly learned how to remove a great many pathogens, toxins, heavy metals, and other dangerous materials from our effluents, although a whole host of contaminants—including hormones, heavy metals, and antibiotics—remained. All the same, the authorities decided cities could reuse the treated water, over and over. We started to recycle paper, also for "green" reasons, which required a de-inking process to remove the chemical inks so that the cleansed paper could be resold—a process that generated toxic sludge as a by-product.

At first, all the dangerous material filtered out of our sewers or removed from paper was landfilled or taken by tanker for deep-sea disposal. But by 1992 marine researchers discovered that municipal and pulp mill sludges were so toxic they were causing enormous "dead zones" along all the major coasts of the world. Sludges were declared to be too poisonous to dump at sea, but they were backing up in landfills and other storage facilities on land, some of which had stopped accepting them, also because of fears of toxicity. What to do? This was the Reagan/Mulroney era, so, by the stroke of a pen, the American EPA renamed sludge "biosolids" (MRF or *matières résiduelles fertilisantes*, "residual fertilizing matter" in French) and declared all of it a valuable fertilizer. The Canadian Food Inspection Agency followed the EPA action quickly, as it almost always does. And since it is illegal to put sludges in water, jurisdictions across North America decided to put it where we grow our food—where more than a third of it almost immediately washes into the surface water, percolates into the ground water, and ends up out in the oceans, after having contaminated agricultural soils along the way.

Our governments could have thought the whole process out a little better and come up with a viable way to deal with water treatment wastes. Closed-system incinerator, gasification, and "plasma incineration" plants are capable of turning even industrial sludge into a harmless, shiny silicon puddle, along with a very small amount of toxic ash that can be sequestered in cement. The more sophisticated plants harvest electricity by burning the biogas emitted by the sludge; one in St. Paul, Minnesota, removes 90 percent of the contaminants and creates $4 million a year in electricity. A few of these plants operate in industrialized countries, even here and there in Canada, but no one has built enough of them to fully deal with the problem. Another solution is to separate human wastes from industrial ones at the source, as the U.K. is doing. Many industrial contaminants like dioxins and PCBs are being phased out anyway, and more should be, with any remaining toxins being carefully landfilled. Still another option, possibly the best, is closed-system reed beds and constructed wetlands. Marsh plants are able to gradually remove toxins and sequester heavy metals in their own cells, which are eventually harvested and incinerated.[31]

The use of sludge on agricultural land is suspected in most of the recent cases of *E. coli* contaminations that have inspired recalls in the U.S., especially in leafy vegetables like lettuce and spinach, which can suck bacteria, viruses, and heavy metals into their cells along with water. Fearing lawsuits, big processing companies like Gerber, Heinz, and Campbell's are now refusing to buy produce from sludged land. Provinces like Quebec reassure the public by saying that only so much municipal or paper mill sludge is allowed per acre per year, but that's like saying that only so much poison is being added to your food supply annually. The deleterious effects, especially from heavy metals, are cumulative.[32]

The good news is that the countries that began this practice before North Americans did are already enacting bans and finding viable alternatives. Switzerland has a ban, and Germany, Denmark, and France are on the way. In the U.S. and Canada, many municipalities have real or *de facto* bans in place. And there are some serious black-water heroes out there. Because most higher jurisdictions in North America have approved sludge spreading, it has been very difficult for any rural area to prevent this material from being dumped on its territory. But in November 2006, the tiny municipality of Elgin, Quebec, composed of only a few hundred souls, passed a bylaw prohibiting the spreading, storage, or transport of municipal or de-inking sludge. They knew they were inviting a lawsuit, but they were particularly worried about the heavy metals in de-inking sludges, which can sterilize soils in a relatively short period, rendering them useless and dangerous for crop production. The fact that sludge could be trucked in from distant, large cities—in their case, from Ottawa—and its provenance could not be monitored was also a big worry. So Elgin passed a bylaw in order to "protect public health and the environment."

The local farm wishing to spread the sludge took Elgin to court. They were backed by a huge multinational waste management company, GSI, which makes its money by taking human and de-inking sludge off the hands of overwhelmed municipalities, calling it a crop fertilizer, and selling it to farmers at much cheaper rates than regular chemical fertilizers. On October 1, 2009, three years after the initial act of defiance, and to almost everyone's surprise, Quebec's Superior Court upheld the town's bylaw. Judge Steve J. Reimnitz invoked the well-known Supreme Court ruling in a similar ban on cosmetic pesticides in Hudson, Quebec. In his thirty-eight-page decision, he also invoked the precautionary principle: "Where there are threats of serious or irreversible damage, lack of full scientific certainty should not be used as a reason for postponing measures to prevent environmental degradation."[33]

What such legislation means is that each one of us has to start taking responsibility for our own town or city's black-water wastes, wherever we live. Continuing to dump our daily secrets out of sight forces the wheel of water to rain those wastes back down on our heads. We have the technologies and methodologies that will keep our water safe—and we must invest in them.

FREE THE RIVER

We can't live without the marshes or without the water... So you can imagine our feelings when the water came back.

A Marsh Arab on the restoration of the Iraqi marshlands[34]

Ever since the early twentieth century, dams have been hailed as the answer to problems of water scarcity in all countries, especially poor ones. Dams reroute water from one area of a watershed to another, control flooding so that people can settle on floodplains, open up more rivers to commerce, and expand agriculture into arid areas. There's a bonus in that the controlled water can be directed through turbines to produce electricity. But there's a big difference between a small catchment dam to preserve rainwater in a dry riverbed and a huge dam affecting an entire watershed.

Small dams don't interfere with the normal flow of a river and are usually able to allow water to percolate down and replenish the water table, gradually restoring the entire system. Until recently, small "run of the river" power-generating dams were considered fairly sustainable. However, we now know that having too many small dams on the same river and rerouting rivers to feed such dams can be deleterious. The free flow of the river is vital to the health of ecosystems along its entire length. And really big dams are usually designed to gain total control over the natural flow of rivers for electricity generation. It's not unusual for whole river basins to be altered and for large populations to be uprooted.

Dammed rivers may change direction and end up draining another watershed; silt that used to fertilize floodplains and estuaries, that provided annual nourishment for crops, is held back; vital plant and animal species that depended on the seasonal arrival of water for their reproduction, hatching, or migration disappear.

As often happens with human developments, we mastered the technological skills to make these massive changes before we understood their effects. We were seduced by the exciting sight of lights going on in valleys and water rushing into dry fields. As Indian author and activist Arundhati Roy says, in most countries the ability to engineer such gigantic structures became closely associated with the whole idea of modernity and progress. Dams were and often still are seen as "a kind of concrete flag of patriotism." They've marched across the major valleys of almost every country in the world, and despite their enormous downsides they are still seen as symbols of a nation's ability to attain a high level of industrialization and technological expertise. Which explains why, although they're now mostly illegal in the First World, they are still being built in places like Africa, South America, and China.

To say people in China or Africa shouldn't have what people in Australia or France enjoy is certainly unfair. The problem is not fairness, however; it's the sad fact that we have since learned the price of progress: nature's bounty is not there anymore. All of humankind has to very quickly evolve a way of life that is not based on having more and more of everything but rather on sharing what is available in order to meet our absolute requirements. It's particularly telling that the original reason for damming the Yangtze, proposed back in the 1930s but finally undertaken in 1993, was to control flooding. But that flooding was largely caused by the deforestation of the river's banks and residential and industrial settlements on its floodplain. The situation would have been a lot easier to control by planting trees and limiting the floodplain to agricultural, not residential or industrial, use. The need for energy could have been fulfilled, albeit more slowly, but with long-term benefits for all, with true renewables and especially with conservation and efficient use. This is true for most of the floodplains of nearly all the dammed rivers of the world. And they still could be managed that way.

It took a couple of generations for the inefficiency and downsides of big dams to be recognized in the U.S., a bit less than one generation in much of India and Southeast Asia, but now opposition is moving at blinding speed. Today there is a robust resistance even in politically repressed China, and protests against the Yangtze's Three Gorges Dam and other large dams are succeeding. In 2004 the Chinese prime minister unexpectedly canceled plans for thirteen dams along the Nu River, a UNESCO Ramsar site and one of the last free-flowing rivers in Asia. A Sichuan dam was canceled after protests about its flooding of a Qin Dynasty waterworks. Chinese scientists and journalists have become increasingly bold about discussing the downsides of hydropower, especially after remarkable statements by Wang Xiaofeng, of China's State Council (the highest executive body in the government) about "environmental security" and about "hidden dangers" of the Three Gorges Dam. "China Warns of Environmental 'Catastrophe' From Three Gorges Dam," proclaimed *Xinhua News*'s English-language website, in reporting his comments in September 2007, astonishing dam opponents.[35]

Today there are increasing numbers of outspoken officials and scientists, like the authors of a report for the city of Zhaotong, where 100,000 people may be displaced by the Xiluodu Dam, who were permitted to publish this opinion: "Past experience has... taught that hydropower development will not necessarily improve local social and economic conditions. There is widespread concern that, although the stations are as modern as those in Europe, the residents will become as poor as people in Africa." In June 2009 China went further, suspending approval for three power stations along the Yangtze when it became clear that "two of the mainland's biggest power companies had begun illegal construction" in an effort to bypass environmental reviews. Another dam on the Nu River was also suspended in May. "What we see now," said Ma Jun, director of the Institute of Public and Environmental Affairs, "reflects a decision made by the very top leadership to balance development with environmental protection." He added, "this is not an easy decision to make in the middle of an economic crisis, and it illustrates Beijing's determination."[36]

Most First World countries have not only stopped building big dams; they're taking down the ones they used to have and freeing the once-captive rivers. Dams have surprisingly short life spans, sometimes as little as twenty years and usually no more than eighty. They crack, and their reservoirs silt up so badly that the water no longer goes through the turbines or flows downstream as intended. Since North American dams were the first to be built, they're the first to have started falling apart. In the U.S. and Canada alone, more than two thousand dams are believed to have outlived their usefulness or are considered unsafe. Today the people who have learned about the issues of climate change, biodiversity, water safety, forests, estuaries, fisheries, and floods associated with dams are demanding that the costs and benefits of every dam be reevaluated in the light of knowledge we have acquired since they were first built.

We're learning very exciting things about the multiple benefits that having a free river brings. When the 50-foot Marmot Dam on the Sandy River in Oregon was removed in 2007, the hydrologists estimated that it would take the river two to five years to pro-cess the silt, rocks, and sand that had collected for decades in its reservoir. The river cleaned itself out in a few months, and "the day after the dam was removed, federally protected Coho salmon were migrating past the former dam site."[37] Only one year after the Edwards Dam was removed from Maine's Kennebec River, millions of alewives, a migratory fish species that depends on upstream spawning grounds, returned to a stretch of the river that hadn't seen an alewife in 160 years. In Wisconsin's Baraboo River the num-ber of different fish species more than doubled, going from eleven all the way up to twenty-four when dams were removed.

Dam deconstruction has even happened in the famous Iraqi marshlands. When Saddam Hussein was in power, he wanted to show off the country's power and technical expertise by rerout-ing this huge wetland's water toward the agricultural north. So he forcibly evicted tens of thousands of the marsh's native people, the Ma'dan—the "Marsh Arabs"—who had carefully managed this ecosystem for thousands of years. The waters were diverted by sys-tems of canals, dams, levees, and a new "river," until the marsh had shrunk to a fraction of its size and most of its inhabitants had fled or been killed. Following Hussein's fall, the United States Agency for International Development (USAID) instigated a marshland resto-ration program. Although only 65 percent of the area has now been reflooded, half is already revegetated, and its native invertebrates, fish, and birds have returned. The flocks of migratory ducks, storks, and hundreds of other species that depended on it as a rare feed-ing ground between Siberia and Africa are coming back, right along with the Ma'dan.

Even the world's really big dams, once immune to any sort of control, are being stopped or deconstructed as well. Turkey was all set to build a monster with a 100-square-mile reservoir on the Tigris, the last free-flowing river in the entire country. But devoted community activists known as Hasankeyf'i Yasatma Girisimi braved a dangerous political climate and got construction stopped after it had already begun. The group received the first Free Rivers Award from International Rivers, the major NGO concerned with dam resistance and river restoration, "for extraordinary success against the construction of the Ilisu Dam... stopped in December 2008."[38]

The International Rivers Network says, "Opponents of large dams do not believe no dam should ever be constructed. They do believe that dams (and other development projects) should be built only after all relevant project information has been made public; the claims of project promoters of the economic, environmental, and social benefits and costs of projects are verified by independent experts; and when the affected people agree the project should be built."[39] Criteria such as these are eminently reasonable. Applying them in North America has exposed the true economic, environmental, and social failures of nearly all big dams, as well as many diversions, dikes, and dredging projects, and we can look forward to the same profound revelations in China, India, and everywhere that the planet's circulatory system extends.

HYDRAULIC CAPITALISM

What happens when you "privatize" something as essential to human survival as water? What happens when you commodify water and say that only those who can come up with the cash to pay the "market price" can have it?

Arundhati Roy[40]

Happily, there are solutions to most of the major problems we're facing regarding the planet's water supply. Active movements and proven technologies are already engaged all around the world in

dealing with scarcity, misuse and waste, alterations of natural drainage patterns, and contamination from infectious agents and poisons.
But the final problem we have with water is perhaps the thorniest:
ownership. Even though water is an absolute necessity for all life and
people have developed many ways to move it around and conserve
it, we haven't really decided who owns it. After all, it moves. Water
tenure is handled differently in different parts of the world, and only
recently have we paid serious attention to which forms of ownership
actually work. Right now the major obstacle standing between managing water in the hopeful ways outlined above, from check dams
and greenbelts to sludge incinerators, is privatization.

There are arguments for privatizing water services that make
sense at first glance or in a few situations. Many studies have shown
that where water is considered a common good to be provided as
cheaply as possible (as it is in most of North America), it's wasted.
And where it's subsidized for agricultural use, as in the American
West, it's wasted on a scale that beggars description. In most of
the EU, especially France, water distribution has been handled by
profit-making companies since the Napoleonic era; France has been
the birthplace for the giants of the industry, like Vivendi and Suez.
The idea is that when people have to pay for water, they will conserve it. Like metered electricity, metered water inculcates habits of
restraint and efficiency. This does seem to hold true in rich countries, where the costs of public metered water go to maintain urban
systems. However, there is no denying that deals between municipal and state governments and big business, for enormous projects
such as water distribution, often go hand in hand with bad service
and political corruption.[41]

About twenty years ago the mayor of Grenoble, one of France's
loveliest and most prosperous cities, made a decision not unlike

the one the Conservative government made in Ontario before the Walkerton tragedy, a decision based on the same philosophy. He decided to privatize the city's water services in order to get "better and more efficient service." Under the aegis of a private corporation, the waterworks in Grenoble would make money for their shareholders; it would also avoid the bungling, overstaffing, and inefficiency of which publicly administered water services are so often accused. COGESE, a subsidiary of Lyonnaise des Eaux, another French water giant, got the contract. As in Walkerton, a lot of people, especially in local government, opposed the privatization scheme, but it found enough political support to proceed.

Water rates immediately soared; services did not. It was later discovered that between 1990 and 1995, tariff increases brought Lyonnaise des Eaux over US$10 million in excess profits for providing water and almost $4 million extra for treating sewage. Even after forced contract renegotiations in 1995, the company still managed to make an excess of $2 million on the city's water and more than $300,000 on the sewage. How did they make so much money delivering a service that was formerly nonprofit? A subsequent trial revealed that during the first six years the company had invoiced customers for 51 percent more water than they actually consumed, which gave them $3 million extra. Then they indexed rates to inflation in such a way that that they were able to make unnecessary price increases of 4 and 5 percent on every gallon of water consumed or sewage treated. As for the idea that pricing water high will inspire conservation, the system set up by Lyonnaise des Eaux granted lower rates to those consuming the largest amounts of water, while those saving water paid premium rates.

As the Grenoblois saw their water services deteriorating, two NGOS, ADES (Association for Democracy, Ecology and Solidarity) and Eau Secours! ("Save Water!"), formed. They led a fight that ended in the 1996 court conviction of Grenoble's mayor and an executive of Lyonnaise des Eaux, for accepting and paying bribes. Not only had the company paid the mayor's campaign expenses, but it was proven that the company recovered the cost of their bribes to the mayor and others, including members of the city council— almost $6.5 million in all—by billing that amount to the water users! Emanuele Lobina wrote in the February 2000 issue of *Focus on the Public Services*, "Corruption is one of the practices adopted by French water multinationals to secure enormous profits. With an increasing body of evidence exposing the irregularities and the costs of the French system of delegated management, this should not be promoted as a global model."[42] Today Grenoble has returned to its tried-and-true system of municipal water management, and water rates are back to normal. The citizens, however, are paying a lot more attention to the issue, and the city has become a leader in pushing for public services across the EU.

The main point of such a story is that publicly controlled, not-for-profit systems are perfectly capable of pumping, piping, and distributing water as well as removing wastewater, and collecting whatever operating fees that these services require. Why should *excess* profits be made out of this necessary exercise? And if a non-profit system doesn't have enough taxes to support itself, how can one that demands profits do so? The answer is, by overcharging users or by denying access to people who can't afford it.

When it comes to expecting private bodies to protect natural ecosystems, the situation is even more problematic. The simple fact is that in the few situations worldwide where water is being managed sustainably, it's being managed by public groups that have to respond to democratic input. Not one of the sustainable methods for conserving water mentioned in this chapter has been developed or is generally favored by the private water corporations of the world, for the simple reason that their focus is on distributing whatever usable water is left on this planet to the people who can afford it. In *Blue Covenant* Maude Barlow points out that "when water is privatized, the public often loses its right to be informed about water quality and standards." Indeed, the Walkerton Inquiry revealed that many more people were sickened while the waterworks managers played around with chlorine and kept the private company tests secret. Under a public system the managers would have been obliged to notify provincial authorities immediately.

In Sydney, Australia, in 1998, people discovered that their water contained high levels of the dangerous parasites *Giardia* and *Cryptosporidium* and that the private provider, Suez-Lyonnaise des Eaux, had known about the contamination for some time but had not informed the public. Terrible problems with access and contamination, including Gabon's first-ever typhoid outbreak, in 2004, led Mali, Gabon, and Guyana to cancel their private water providers. Riots and sabotage have accompanied the installation of prepaid meter systems beyond the reach of the majority of citizens in South Africa and Namibia and have inspired the creation of more of the anti-water-privatization citizens' groups that are spreading around the world.

There are many other examples, but probably the worst are the water privatizers in rural Africa and India who are making deals with government elites to "own" entire watersheds. This means they are charging people monthly fees for taking water from surrounding rivers and streams and other polluted surface water, as well as for groundwater that is delivered by pumps and pipes. Some have gone so far as to charge people for the rainwater catchment systems they have built in their own villages, like cisterns on roofs, claiming that owning all the water in a watershed means they get to charge money for every raindrop that falls. Clearly neither humanity nor the planet can survive if we don't limit corporate power in this arena.

The good news is that public water utilities are increasing again. One example that addresses the entire question of service efficiency is the story of the Servicio Autonomo Nacional de Acueductos y Alcantarillados (SANAA) of Honduras. SANAA was created in 1961 as the state's water supply and sewer service. It was badly set up, too centralized, top-heavy with staff, and, as with many large bureaucracies, it suffered from poor communication between departments. Wages were low, unions were angry, morale was down, and customers were dissatisfied. In 1994 the now-familiar recipe for alleged salvation came down from the Inter-American Development Bank: privatize all your water.

This ultimatum galvanized the Honduran bureaucracy; with union support they opted instead for complete reorganization. Not only would they fix themselves, but they would also aim for a very lofty vision of their collective future. SANAA encouraged its employees to praise each other with buzzwords like "dedication, integrity, pride, and unity" and fostered self-management and self-organization on every possible level. They decentralized, reduced overstaffing by 35 percent, made billing local (a huge improvement), and, while increasing the price of water, made sure that there was a subsidy that delivered the first 20 liters (5 gallons) daily for free, which was enough to keep people without a cent from misery. They then increased pipeline networks threefold and enabled the population to receive piped water twenty-four hours a day, for the first time in their country's history. By making infrastructure repairs that reduced leaks, they were able, in the capital city of Tegucigalpa alone, to save an incredible 26 gallons *a second*. The upshot of their efforts was not only the salvation of publicly-owned water and sewage in their country but the recognition of SANAA, in 1999, as a United Nations Model Project.[43]

Obviously not every private water service behaves as badly as Lyonnaise des Eaux did in Grenoble, and not every public one is as inspiring as SANAA. The point is that there is no empirical evidence that public sector water utilities are less efficient than privatized water companies. Since public utilities don't have to come up with profits to investors, figures show that they deliver the water at a better price. Public utilities can also exist in many different forms. They can raise investment finance like a business, and from all the same sources private companies do—banks, government funds, international institutions like the World Bank or the Asian Development Bank, and bond markets. As nonprofits they can also qualify for grants and aid; the city of Lodz, Poland, for example, found that grants from the Polish National Fund for Environmental Protection and Water Management would finance their new sewage plant much more cheaply than a deal offered them by Vivendi. Water utilities can also be cooperatives, the biggest being SAGUAPAC, in the city of Santa Cruz in Bolivia. And many places are reaping the benefits of a large system of public-public partnerships (PUPs), which arranges for a publicly owned water company in a developed country to provide expertise and help finance infrastructure for a new publicly owned company in a developing country. There are few if any such agreements or initiatives in the private sector. Why would there be? They wouldn't increase profits.[44]

The first-ever pan-European civil society coalition against water privatization was launched in Malmo, Sweden, during the European Social Forum in 2008 and is insisting upon "major changes in

EU policies towards water management, away from the current pro-privatisation approach."[45] This is to address European Commission legislation that requires "competitive bidding" for water contracts but actually results in unrealistic bids and then cost overruns or poor service from the private corporations that have simply pretended to be more efficient than the public ones.

Even in France privatization is crumbling under the onslaught of reality; in 2008 the mayor of Paris ended the city's decades-old contracts with two private operators and is developing an integrated public water and sanitation system. Paris will join Grenoble, Amsterdam, Seville, and Vienna in developing "highly effective, environmentally sustainable, and socially responsible public water management." According to British expert Dexter Whitfield of the European Services Strategy Unit, a pro-public services think tank, the financial collapse of 2008 "demolishes the argument that the private sector is better at managing risk, as governments come to the rescue of the financial sector... Private sector borrowing costs have shot up, making private-public partnerships even more expensive and increasing the advantage of public financing."[46]

Environmental lawyer Jim Olson, involved in a fight against Nestle's attempt to take over Michigan's aquifers for its bottling plants, says that we must learn and repeat the refrain that the privatization of water is incompatible with the *nature* of water. "Water is always moving unless there is human intervention. Intervention is the right to use, not own and privatize to the exclusion of others." Olson warns that if politicians side with the World Bank, NAFTA, and WTO view that countries should surrender rights to their water to for-profit corporations, "that state will have violated the rights of its citizens who have redress under the principle of human rights" under a future world Water Covenant.[47]

Back in 1999 the world found out what would happen when water moving through an area was lost to people through privatization. In Bolivia, then a country ruled by a tiny elite and very much indebted to the IMF, all of the country's water rights were awarded to a profit-making American corporation, Bechtel. The price of this necessity immediately skyrocketed and became too expensive for poor people to buy. Subsequent protests were led by Oscar Olivera of Cochabamba, a humble, shy man, who worked in a shoe factory. His group demanded that the government tear up the privatization contract. In the ensuing demonstrations and blockades 36 Bolivians were killed by police, 175 were injured, and 2 children were blinded.

Today, the world's first indigenous president, Evo Morales, has been put into office largely by the people who organized around the water issue; Olivera has been honored with the Right Livelihood Award; the people of his hometown, where the water-grab first took place, now manage their own water utility; and the "right to water" has been enshrined in the country's new constitution. Bolivians have had to tolerate the seizure of minerals, animals, forests, farmland, and many other resources, by outsiders or elites they could not control. This grassroots resistance to the seizure of water shows there are limits to what people can stand. President Morales has also called for a "South American convention for human rights and access for all living beings to water" that would make his country's rejection of the commodification model for water a new international goal.[48]

Bolivia's position has spread, peaceably, around the world. In March 2009, at the fifth World Water Forum, delegations from Bolivia, Uruguay, Spain, Guatemala, Ecuador, Cuba, and Chile, with support from Bangladesh, Benin, Chad, Ethiopia, Honduras,

Morocco, Namibia, Niger, Panama, Venezuela, Sri Lanka, Switzerland, and South Africa, demanded that governments around the world recognize that "access to water and sanitation is a human right."[49] In August 2006 the Indian Supreme Court ruled that, "the protection of natural lakes and ponds is akin to honoring the right to life—the most fundamental right of all."[50] In water-rich Canada, Maude Barlow's Council of Canadians, along with a new coalition called Friends of the Right to Water, continues to pressure Canada's recalcitrant government on this issue, as part of the global campaign to recognize access to water as a basic human right.

THE WORK AGENDA

When we enter a case, Riverkeeper assembles a team of experts in law, engineering, biology, hydrology, economics, and energy policy. Combining fund raising... use of pro-bono services and keeping overhead to a minimum, Riverkeeper has proven to be the David who successfully topples many a polluting Goliath.
Riverkeeper website[51]

When it comes to water, we have lots of work to do. It's indescribably alarming to realize how profoundly human activity has affected this basic element. Since the first edition of this book was written, many legislative milestones have been achieved, and hundreds of new groups dedicated to the protection of water in all its forms have sprung up in civil society, mostly in response to direct threats. There are also many new sections of older NGOs now becoming committed to water projects. Most of these organizations are run by the usual suspects: local families, farmers, fishers, and so forth; but some have the benefit of skilled directors with international contacts, like Rajendra Singh, now of India's Central Ground Water

Board, Tony Clarke of the Polaris Institute, or Maude Barlow—leaders who are rapidly growing more famous in the defense of water.

Another example is Robert F. Kennedy Jr., an environmental lawyer who back in the 1990s decided to do something about New York City's steadily degrading water supply. He founded River- keeper, which enlisted volunteer boat owners and fishers to patrol the origin of the city's water, the Hudson River, identifying sources of contamination and then prosecuting the polluters. Riverkeeper negotiated the New York City watershed agreement on behalf of environmentalists and the city's water consumers, which is today "regarded as an international model in stakeholder consensus negotiations and sustainable development." They expanded to the Chesapeake Bay area and are credited with halting the most egregious practices of industrial hog farming in that area, which led to a ban on this truly terrible industry in North Carolina.

Riverkeeper then went viral, turning into the Waterkeeper Alliance, which empowers more river users worldwide to take charge of their watersheds, to monitor and protect them. The group has trademarked the names "Riverkeeper," "Lakekeeper," "Baykeeper," and "Coastkeeper" to keep corporations and polluters from capturing these titles and what they represent. Today there are nearly two hundred waterkeepers around the world, patrolling the Atchafalaya Basin in Louisiana, the Thames in England, the Bagmati in Kathmandu, the Baltic Sea and Lake Baikal in Russia, the Middle Han in China, the Bogotá River in Colombia, the Bay of Fundy in New Brunswick, and the Fraser in British Columbia. There are another half a dozen keepers in Australia and close to a dozen in India, where the keepers of the horrifically polluted Yamuna that feeds New Delhi, as well as the lower reaches of the Ganges, have their work cut out for them. There are scores more patrolled rivers in the U.S. and Canada, as well as in all of South America, with new Waterkeeper chapters being started every day.

There is no social crisis so immediate as a lack of access to clean water, and the relationship of the hydrological cycle to climate change is adding to the urgency of this issue. Big, well-heeled organizations like the Global Resource Action Center for the Environment, the Pew Charitable Trusts, Greenpeace, Sierra Club, World Wildlife Fund, and Friends of the Earth are all moving more clearly into water issues. They've found they have to take on these issues whether they want to or not, as they notice the effects of the water crisis on their area of concern—be it forests, farming, or fish.

If we look carefully at the stories told in this chapter, from Rajendra Singh's consensus building in support of an ancient water storage method, to the positive effects of dam decommissioning and the success of Honduras's public water utility, we can clearly see they all have several things in common: the solutions they offer are simple; they are historically well established and scientifically based; and the key to achieving them can be summed up in Barlow's advice: "The best advocates for water are local communities and citizens," who need to "participate as equal partners with government."[52] These solutions also fit perfectly into our criteria for sustainability: they mimic nature, they have lofty goals, they are democratically and locally based, and they consider the needs of ecosystem reality ahead of the endless desires of our fantasy-based economy.

As Victor Munnik, a South African water policy specialist, puts it, "We have to adapt our economic development according to the water resources that we have, [not vice versa.] The impact on our water needs to be integrated from the very beginning into *every* economic development plan."[53] Politicians are clearly a long way away from that. We can't afford to let them stay that way. The long list of organizations above—and there are so many more, in even the smallest town near you—are easy to join. They don't just want your money; they need members, volunteers, supporters, letter writers, voters, and petition signers and distributors. They need people with boats and people who appreciate lakes and rivers, ocean beaches and spring rain. They need people who depend on water to survive. Oh, right—that's all of us.

6

THE MOTHER OF ALL

OCEANS

SAVING THE SEAS

Some scientists I have met argue that instead of calling this the age of "global climate change," we should call it the era of "global ocean change" or "marine climate change."

Alanna Mitchell[1]

We know that life in all its forms began in the sea and that the water cycle—the continuous movement of water on, above, and below the surface of our planet—begins there, too. Even after life-forms evolved to become land based, most of them, plants and animals alike, could only survive away from their mother's embrace by carrying their own stock of salty waters within their bodies. Surprisingly, researchers are just now turning their attention to this planet's marine realm—its chemistry; its powers of absorption, suspension, and dissolution; the fluctuations in its temperature and currents; the marvels of its mountains and canyons; and especially the weird and wonderful universe of the incredibly bizarre life-forms lurking in its greatest depths. The world's first Census of Marine Life, begun less than ten years ago, doubled the number of known marine species in its first three years of research alone.

Paying real attention to the oceans is so recent that the first international conference to seriously address the intersection of climates and oceans wasn't held until 2008. Author Alanna Mitchell notes, "the first scientific attempt to map the human effect on the world's oceans [only] came out in the journal *Science* in February 2008." There are few things more important than reading Mitchell's book, *Sea Sick: The Global Ocean in Crisis*. It's the first mass-audience book that attempts to explain what all human activities, especially energy consumption and industrial farming, have been doing to the oceans in terms of pollution and fisheries depletion and also in terms of the warming and acidification of seawater caused by our vast production of greenhouse gases.

We often pay lip service to the biological richness of the oceans and the effects upon them of the modern fishing industry's wholesale liquidation of life. But few of us realize how devastating destroying all the wild fish in the sea actually is, not just to the oceans but to the entire planet's stability. Mitchell makes it clear that the oceans of the Earth are "a biological gold mine that is even more productive than tropical rainforests; arguably the most important medium of life on the planet." A staggering 99 percent of all the living space on this planet is water, and oceans are home to at least half of Earth's mass of life. The main life-form in oceans is plankton, the tiny organisms that are the base of the food chain, not only for sharks and whales but for humans as well. As Mitchell points out, "Plankton produce half the oxygen we breathe." She notes that until the last few generations, although we dimly realized that the oceans are everyone's true home, grocery store, and sewage system, we hadn't realized that they are "our main life-support system, controlling the planet's temperature, climate and key chemical cycles."[2]

We are so generally ignorant of this life-supporting activity that some proponents of geoengineering schemes to sequester carbon want to inject our excess CO_2 into the sea, ignoring the fact that carbon dioxide in water is active. As the oceans try to deal with human overproduction of this gas, seawater is becoming more acidic, and its fundamental chemical composition is changing. With every plane ride we take, the oceans are becoming toxic to life. At the same time "dead zones" are spreading off every coast. They're caused by pollution, largely from agricultural runoff but also from human and industrial wastes. There are hundreds of these huge blobs of low-oxygen water, many of them as large as a small American state. They "are thickening and moving nearer to the surface, the direct result of the climate change that is warming the ocean. In turn that keeps the colder, even deeper waters—which contain all the food and nutrients necessary for plankton—segregated from the surface."

When we wrote about oceans in the first edition of this book, we thought that all people had to worry about was overfishing with longlines and trawlers plus massive pollution from farming and dumping. But added to these scourges is the newly defined problem of acidification, caused by the ocean's attempt to absorb humanity's enormous production of carbon dioxide. Seawater is becoming so altered that in places its acidity not only kills live coral but also dissolves the beautiful, multifarious seashells that ocean creatures create from calcium to protect themselves. In fact 80 percent of the heat that has resulted from global warming has not warmed up the land or the air but has been absorbed by the oceans. The last time the oceans' temperature and chemical balance were changed on a scale similar to what is happening now was during the Permian extinction, 250 million years ago.

The size of this problem is obviously terrifying. Even in the rare cases where ecosystems are being protected, say the Great Barrier Reef off Australia, there is no way to protect marine habitat from changes in temperature and acidity. Scientists worry about feedback effects on land—for example, when forests become dry and pest infested from warmer winters and the resulting forest fires cause more carbon dioxide emissions and more warming—now we're looking at a feedback loop at sea. There will be, as Mitchell puts it, a chemical flip, a point of no return. "The evidence is that the chemical levels of its ocean-blood are changing and that is affecting such things as pH, metabolism, fecundity and ability to thrive. In a growing number of places, the very oxygen content of the ocean is trailing off. These are its vital signs, and they are telling us that the planet is... slipping into biological and evolutionary unconsciousness."[3] That doesn't mean all life will stop, but these changes do mean far less diversity. New systems will come into being, as they already have in the Caribbean, but humans probably won't like them. There, the brilliant coral forests swarming with fish, crustaceans, and sea mammals have been replaced by dead coral bones smothered in miles of slimy algae.[4]

Good news about oceans is in short supply right now. Unlike energy or agriculture, where sustainable solutions are already understood and starting to be globally applied, the human response to the oceans crisis isn't nearly as advanced. Making the connection between the land, the sea, and ourselves will help us get a handle on this monster of a challenge. By and large, people don't live on the oceans, so we can't clean up our act out there except by outlawing pollution, dumping, and overfishing—and of course we need to get very busy on that. But most of problems are coming from the land and our behavior there. If we deal with all forms of pollution, we will also be addressing ocean acidification—very quickly and directly. If we reform agriculture, which is already happening, if we protect our forests and wildlife and rivers and watersheds, if we outlaw plastic waste and learn new values that are based on healthy systems, not fantasy money—same thing. The simple fact is, all things eventually end up down in the sea. That means that every sustainable solution discussed in this book will help the Mother of All to survive and keep the great wheel of water turning.

STRIPED GLORY

You could fish for bass at the best areas all night at the best time—the full moon in October—and not get a single bite... In my opinion, this recovery was the greatest success story in fisheries management in the world.

Carl Safina, Audubon Society's Living Oceans program[5]

Any avid fisherman knows that one of the great experiences in angling is to stand in the spray of cold surf on the Atlantic coast, cast plugs into ravenous schools of striped bass, and finally, after a glorious battle, haul in a forty-pound beauty. For sport fishers, striped bass held the same exalted position on the east coast of North America as chinook salmon have on the west. Like salmon, they are anadromous, dividing their life cycle between fresh and salt water, and have dense, fatty, and delicious flesh. In increasingly populous and industrialized U.S. coastal states like New Jersey, New York, Massachusetts, Rhode Island, Maryland, and North Carolina, all species of wild fish have had to contend with human predation by sport and commercial fishers, deforestation, effluents that have been dumped into the rivers and that have interfered with

reproduction, and more. Nevertheless striped bass remained plentiful until the 1970s. After that their numbers dropped precipitously
year by year.

Devastated sport fishers, environmentalists, and governments,
as well as all the industries that depended on this resource, began
pointing fingers. Commercial and recreational fishers blamed one
another for being greedy. Water quality, pollution, and urban development were cited as culprits. All these factors played a role, but
the real problem was the fact that the striped bass had been treated
as a cog in the wheels of a state economy, not as part of an intricate
natural system with its own complex and ultimately very restric-
tive regulations.

Resource experts and managers had assumed that the survival
of young bass (West Coast managers have assumed the same thing
for young salmon) is *not* dependent on the numbers of spawning
adults. In their opinion the only factors crucial to the size of the
bass population were the conditions surrounding the eggs and
fry—water quality, temperature, turbidity, and so on. Even though
it would seem logical to assume that if more mature fish are lay-
ing eggs, more juveniles will survive, the fry were monitored but
the spawning adults were not. State resource agencies permitted
so much overfishing of the adults that not enough of them were
surviving to breed. In other words, the state assumed that the fish
would somehow manage to reproduce at a rate commensurate with
all the stakeholders' desires to exploit them, even if they were killed
before sexual maturity. This assumption led government managers
to overestimate the sustainable catch, just as they did with Atlantic cod in the 1980s and are still doing with Pacific salmon. It's a
management style that really doesn't take into consideration the
rather obvious fact that fish need to grow to maturity in order to
have offspring.

Eventually all the stakeholders realized that although the decline in striped bass was a result of complex factors, the only one that was immediately controllable was human predation. Since the Chesapeake River produced 90 percent of the striped bass on the coast, in 1982 the United States Congress passed a federal emergency plan to protect the juveniles surviving in it; they limited human activity, a prime requirement of adaptive management. Fishers, both sport and commercial, were only allowed to keep bass above a certain minimum size, in order to ensure that the new batch of juveniles would have a chance to mature and spawn. Many state governments and fishing industries objected to this interference, but the federal officials forced everyone to comply with the new rules. They also began a massive publicity campaign to enlist the support of the fishers, offering the hope that the striped bass might one day return to delight the hearts of anglers. Throughout the 1980s it was impossible to go into any store on the East Coast that sold fishing tackle, licenses, or even bait without seeing pictures of striped bass and reading the notices explaining the heroic sacrifices people would have to make to bring them back.

Striped bass are predators high on the food chain, so contaminants can become concentrated in their flesh as they feed on smaller organisms, a process termed "biomagnification." High levels of PCBs were discovered in striped bass in the mid-1980s. So there was yet another reason to curtail the fishery; recreational and commercial fishing on the Hudson River were both closed. Although the states bordering the Chesapeake River protested, this closure quickly resulted in increases in the number of fish, the ironic result of the fact that their flesh was too contaminated for humans to eat. Acceptable size limits of striped bass along the coast were gradually raised over the next few years, allowing more fish to reach maturity. Commercial fishing was strictly controlled, while recreational fishers, whose catch had once been unlimited, were allowed only one fish a day.

On a Cape Cod beach one hot, sunny July day in the late 1980s, Holly was asked by two excited surf fishermen who'd hooked a big striped bass to take a picture with her camera and mail it to them. When she asked them if they planned to eat it, they looked scandalized and explained that this was a striped bass; they had almost become extinct. All they wanted was the excitement of the catch. After the photo, all three watched it swim away. That bass was not the only one to be spared. Statistics show that sport fishers not only kept within the quota but did much more, releasing nearly *all* the striped bass they caught directly, settling merely for the thrill of the tussle.

So, within a decade, larger numbers of spawning adult striped bass returned to the rivers, more eggs were laid, and the numbers of juvenile survivors rose steeply, disproving the old management assumptions. By the mid-1990s, the whole length of the East Coast had bass again; they rebounded so strongly that, as marine scientist Carl Safina says, "You could catch numbers of large striped bass in the middle of a summer day, which was previously unheard of. The whole idea that led to their return was protect the fish by size limits designed to allow each female to lay eggs at least twice, and leave enough fish in the water in general, by restricting catch limits." In other words, manage adaptively: control human activities in order to respect the needs of the fish.

Today, there has been some backsliding. As soon as the fish rebounded, there was enormous pressure from industry and state governments, especially New Jersey's, to open the fishery again— enough pressure to get government officials to reduce the size limit to 28 inches and double the sport bag limit to two fish per day. That pressure continues today. The result is that average sizes and numbers have noticeably declined and fish over thirty pounds are rare again. There are no more posters about striped bass in the bait shops, and users, especially nonlocal sports fishers vacationing on the coast, can be forgiven for thinking that if there were a crisis, they would be told. Stakeholders' immediate and enthusiastic cooperation when the resource was perceived to be in trouble gives some insight into the way people are willing to limit their own activities *if they are well informed about its real levels of resilience.* We have done too little to keep users informed and trusted too little in their self-control.

It's become a rule in most conservation circles that the way to sustain species on this planet is to allow local users to manage resources on a day-to-day basis but never to give power to intermediate bodies such as provinces or states, because they are very susceptible to economic pressures. Larger bodies, like federal or international agencies, can establish ground rules and standards and, if necessary, enforce compliance. But that enforcement will never work without the understanding and approval of local users. When the users who depend upon a resource are truly local, the picture gets even better. They are the best source of knowledge and make the best long-term managers we are likely to find.

COMMUNITY FISHING

I'm a fourth-generation commercial fisherman. I'd love to see my son get involved in this commercial fishery.

Wes Erickson, halibut boat captain, British Columbia[6]

Halibut can grow to 250 pounds or more. They bring a premium price and are a prize target for commercial fleets. Until recently the

Canadian Department of Fisheries and Oceans (DFO) would determine in advance which boats would go after which species, and then allocate maximum limits on the weight of fish that, for example, a halibut fleet was allowed to take. It would also allocate a season to catch—only a week or two. Anxious to get as many as possible in this small window, the boats used longlines baited with thousands of hooks, laid down on the ocean floor hundreds of feet below the surface. When the lines were pulled up, many other species of hungry fish would have taken the bait, like red snapper, sablefish, ling cod, skate, and sharks. These fish are called "bycatch"—other boats would have had licenses for them, but the halibut boats wouldn't. They had to throw these fish away. Stories abound of boats licensed for halibut or salmon leaving a trail of floating, dead snappers and sharks in their wake. Because the bycatch was often killed by the hooks, some of it was kept and sold illegally, but that was not as much of a concern as the fact that this management method clearly had some big problems.

Today those problems are being addressed. At first the DFO wanted to prevent people from keeping the bycatch, but that approach didn't deal with killing too many fish in the first place. Human observers on each boat were too costly and of course might be bribed. So the department decided to let the fishers themselves come up with a better management strategy. That's when the real changes came.

The fishers asked for, and got, a much longer season, over months rather than weeks, in which they could exchange licenses for types of fish between boats. In other words, a halibut boat catching a large number of rockfish could exchange limits with a rockfish boat that had caught halibut; each boat would more quickly reach their limits, and this process would eliminate waste. The fishers also asked for two tamperproof video cameras on their boats that would turn on any time equipment on the deck moved. The data could be compared with the skipper's log book and the sale records and would help keep track of bycatch. Moreover, the cameras could provide scientists with a far more accurate picture of fish populations.

Although the urgent issue of using longlines at all still needs to be addressed, these changes are helping scientists learn how many fish are there, so it's easier to figure out a limit. The users of the fish, like halibut boat captain Wes Erickson, are the people with the most to lose in terms of management decisions. Including them in the management process is the most important step toward sustainable fish management, toward figuring out what's really in the sea and how much restraint humans have to exercise to keep the resource going.

A good example of restraint can be found in Louisiana, where some well-established, close-knit community users and their decision to use minimally invasive oyster harvesting technologies, like hand tongs, have made the state's oyster beds on the Gulf of Mexico a model of sustainable use.[7] These oystermen, like the halibut fishers, have managed to gain a form of secure tenure over a particular resource. In general, shellfish harvesting is a form of fishing that easily bleeds into farming; because the resource doesn't move, it's easier to "own" it. Commercially maintained oyster and clam beds are one of the most sustainable types of fisheries because there can be local commitment and clear control over the resource. However, other kinds of fish farming cannot be viewed in such a positive light.

Like biofuels, fish farming is one of the best-intentioned but most awful ideas people have ever had. It makes sense in theory but in practice turns out to disrupt delicately balanced marine ecosystems on every level. Besides being wasteful (salmon farms use

edible fish like herring and sardines as feed, thus removing protein needed by other life-forms), the fish farms that have grown
up along coasts, from Nova Scotia to British Columbia, Chile to
Thailand, have been environmental disasters. The species in the
cages are often alien and can affect local ecosystems when they
escape (and they always do), while antibiotics used to keep the fish
healthy pollute the waters. But most seriously, diseases and parasites infecting the farmed fish can explode because the animals are
packed together so closely, just like industrial hogs. Sea lice are the
biggest problem salmon farmers face and are a major cause of the
collapse of the wild salmon fisheries around the world. These parasites come out of the crowded farms and infect wild young salmon
just going out to sea, gradually weakening and killing them before
they can reproduce.

Putting fish farms in tanks on land is helpful but expensive,
especially as currently, large fish-farming companies can dump
their environmental costs on the local ecosystem. There is some
possible good news in the work of Thierry Chopin, a seaweed
expert from the University of New Brunswick, and Shawn Robinson, a biologist with the DFO, who have been trying to revive an
ancient Chinese practice that eliminates some of these problems
biologically. They approached Glenn Cooke of Cooke Aquaculture,
a big fish-farming company in the Maritimes, and began their program of Integrated Multi-Trophic Aquaculture (IMTA). The waste
that flows out of the open-net pens of fish farms is a huge problem,
so Chopin and Robinson tried growing mussels, which love to eat
waste, around the outside. They also encouraged seaweed growth
beneath the nets. When the mussels were tested for health after-
ward, they passed so well that both they and the seaweed were
found to be marketable.

Another problem with fish farming is caused by overfeeding. It's hard to tell from above how much the fish eat, and many pellets end up falling to the bottom of the nets and creating mounds of rotting waste. Cooke had cameras installed below every net pen so that as soon as pellets start to drift down, the technicians know the fish are full and stop feeding. That saves money and reduces pollution. As for what's left, sea worms, cucumbers and urchins are being grown to mop it up.

Using natural species this way means, as Chopin says, "we have biofilters, nutrient scrubbers, but at the same time, they have a market value." By using more plant material, Cooke has been able to reduce the amount of wild fish in their feed to less than one pound for each pound of salmon produced, so this experiment is encouraging on almost every front. They are attempting to address the biggest problem of all, sea lice, by growing far fewer salmon per net. So far it's only an improvement, not a solution, but at least Cooke is trying. The company doesn't grow species that are alien to the area, and it uses genetic material from wild salmon for their "crop," so that if any escape the impact will be lower. These are all steps that could lead toward a more sustainable form of ocean farming, if one is possible. The still-faraway goal would be finding a way to get rid of the lice—and then passing legislation that would make sure that IMTA is practiced throughout the industry.

OCEAN CORES AND CORRIDORS

It might not mean less fishing, but it does mean you don't fish everywhere.
Bill Henwood, Parks Canada[8]

For centuries, "the deep blue sea" beyond the continental shelves was considered almost devoid of life, while huge sections of ocean out in areas once called "the doldrums" were viewed as complete deserts. Now we know that life is merely layered below, or that it may take a form we did not recognize. For example, sailors have routinely avoided a region in the North Atlantic Ocean known as the Sargasso Sea, part of a windless gyre filled with floating sea-weeds and surrounded by four major currents, including the Gulf Stream. Besides being key to the survival of European and American eels and young loggerhead turtles, it turns out to be filled with huge gelatinous organisms called salps, which can be six feet across. Salps are filter feeders—so efficient at filtering nutrients from water that their organs can be used to filter viruses! They live on nano-plankton barely larger than a molecule, but the waste they excrete after consuming their invisible meals is the size of a mouse turd. Filmmaker and marine biologist Blad Hansen says that means the carbon in the plankton that these creatures have gathered from the surface of the ocean is being sequestered into the deep sea, and at enormous rates. They are performing a key ecosystem service, and over the past few years we've only scratched the surface of their activities.

We know so little about the creatures that comprise so much of the biomass of the sea because their watery bodies collapse when we bring them up from the depths and out into the air. Now that we can film them in their natural habitats, we're beginning to appreciate the huge part they play in supporting ocean life, rather as if we were ancient people stumbling upon the buffalo herds of the Great Plains after years of seeing only birds and ground squirrels. Salps and jellies are just one group we're learning about. For example, most of us grew up thinking of octopi as gray or black with pinkish tentacles. But now we know that there are hundreds and hundreds of varieties: purple, pink, and striped ones; small, ringed, or polka-dotted charmers that are extremely poisonous; even the (no other word but "adorable" will do) "Dumbo octopus," a deep-sea variety with a pair of fins that protrude near the head like giant elephant ears.

Besides the extremophile life-forms living near undersea volcanic vents; bacteria; tiny white crabs; and incredibly beautiful, feathery tubeworms, there are "gulper eels," which can unhinge their jaws to engulf prey several times their size; monsters that seem to be nothing but nightmares of teeth; and a plethora of jelly beings festooned with colored lights and flashers that are too weird and wonderful to be believed. Even familiar species we've eaten for years, like rockfish and red snapper, we're learning can live to be 120 years old and have been genetically selected to occupy certain levels of the ocean. That means when a community is fished out, it cannot be recruited and reformed from outside. Usually we don't know enough about the basic biology of *any* ocean species to be able to manage it.

Fortunately, however alien-looking the life-forms, this is still Earth, and the same general physical laws and conservation policies that work on land apply to the oceans. The 4Cs of Chapter 2—Cores, Corridors, Carnivores, and Communities—work as well underwater as in a boreal forest. Despite scientific recognition of this fact, however, the number of national and international preserves in the seas, or Marine Protected Areas (MPAs), is still less than 0.7 percent of the oceans' area. This paltry effort at conservation must increase immediately. That can happen with political push and citizen pressure, but the most immediate action we can all take is at the grocery store, using guides put out by online fish providers like Eco-Fish, standards organizations like the Marine Stewardship Council, FishWise, and Seafood Safe, which also tests for the buildup of toxins like mercury. Organizations like Greenpeace International, the Environmental Defense Fund, Sierra Club of Canada, and even the Monterey Bay Aquarium also offer consumer guides.

In Canada, an alliance of environmental groups, including the DSF, has worked with chefs in our large cities, encouraging them to cook only sustainable seafood and to do so with pride. Called Sea Choice, this program sets up three standards for consumers, from red (avoid) to yellow (some concerns) to green (best choice). These guidelines are online and suggest that consumers limit seafood dinners to small shrimp, mahi-mahi, wild Pacific salmon, Pacific halibut, pollock, mackerel, striped bass, and other sustainably caught species. The red list tells shoppers which species to boycott, including all sharks, orange roughy, all kinds of tuna except skipjack, as well as all rockfish, monkfish, marlin, and Atlantic halibut. The guides also tell consumers which aquaculture programs are safe (bay scallops, rainbow trout) and which aren't (tiger shrimp, Russian caviar, and so far, *all* farmed salmon). Sea Choice has been adopted by Overwaitea, a chain of food stores in Western Canada. Unlike chains that offer organic products on the same shelf as less sustainable choices, Overwaitea has committed to protecting marine species at risk by not selling them at all—a commitment that needs to be expanded to other chains.

Like the logging boycotts that have created standards and saved so many forests, this campaign to shop and eat sustainably is very much the start of something big. Governments know how to restrain and punish poachers on land and are just beginning to respond to pressure to start doing the same thing on the sea. The first line of defense is these guides, which help destroy the market for species being ruthlessly slaughtered. The dredgers, trawlers, and longline vessels are often operated by Mafia-like organizations like the Russian mob and the Japanese Yakuza. The latter, for example, are deeply involved in the continued hunting of sharks for their fins and whales in protected waters.[9] The next step after that is

implementing the ocean equivalent of the 4cs by truly protecting
species in a wide variety of marine habitats.

Research released by the National Center for Ecological Analysis
and Synthesis in the United States found that the average population density of marine life is an incredible 91 percent higher within
the world's few protected marine areas—the "Cores"—than it is
outside them. Moreover, the actual size of the marine creatures
is 31 percent greater, and species diversity is 23 percent higher. A
book by Callum Roberts and Julie Hawkins, *Fully Protected Marine
Reserves: A Guide*,[10] notes that even a tiny reserve like the Hol Chan
in Belize can have an amazing impact. It was established in 1987 as
a no-take zone of only 1 square mile to address some of the damage
of overfishing and destruction of the shore mangroves. Four years
later this little haven from human predators had produced a six-fold
increase in biomass of commercially important reef fish. The De
Hoop MPA in South Africa, which covers 32 miles of shoreline and
extends a mere three nautical miles out to sea, protects sixty of the
most heavily exploited species in the area, victims of recreational
and commercial line fishing as well as beach seining and trawling.
Their numbers have bounced up within the park *ten-fold* since the
reserve was instituted in 1985, and of course since the fish swim
around, the commercial and sports fisheries outside the MPA are
also benefiting enormously. Kenya is a shining star in the MPA liter-
ature, and in the eighteen years since it established the very modest
Mombasa Marine National Park, covering only 4 square miles, fish
biomass within the park grew more than five times greater than in
the exterior, commercially fished areas. This park has had to institute night patrols to prevent poaching.

There are many more examples all showing similar effects.
Larry Pynn, author of a 2001 *Vancouver Sun* series on MPAs, wrote
that "No one is saying that marine reserves are the only answer to

long-standing problems that plague fish stocks... But the over-whelming body of evidence is that marine reserves—not just one or two, but a network encompassing all ecosystems—are an integral part of fisheries management."[11] There's plenty of controversy about whether they work best when totally protected—which tends to increase large fish and decrease middle-size ones—or whether some commercial activity can be allowed so long as nursery areas and other prime habitat aren't disturbed. A consensus statement signed early in 2001 by 161 leading marine scientists concluded that "Marine reserves are beneficial for conservation and biodiversity and... enough information exists to justify *their immediate creation.*"[12]

One challenge, of course, is to understand the oceans well enough to know which parts to protect. Besides the well-known marine habitats like reefs, kelp forests, and mangroves, there are muddy lagoons, sea-grass beds, river estuaries, sand bars, islands, inter-reef gardens, and the saline-to-brackish-to-freshwater marshes that line many coasts. They're all part of the ocean's nurs-eries. Many marine species, like salmon and striped bass, require several places to live out their life cycle; preserving one area because adults live there is not going to ensure that the fry have their own habitat in which to grow, and vice versa. For example, red emperor fish larvae hatch from eggs laid along outer reefs; they drift in with the tides to inshore nurseries and only return to the outer reefs as adults. Jon Day, director of conservation, biodiversity, and world heritage for the marine park authority of the Great Barrier Reef in northeastern Australia, says, "A lot of these bioregions are not sexy, but if we don't protect them, then we're just kidding ourselves that we're looking after the coral reefs."

Looking after any "Core" area also means looking after the "Cor-ridors" between them. Canada's most aggressive NGO in this field, the British Columbia chapter of the Canadian Parks and Wilderness

Society (CPAWS), has been pushing for a network of reserves similar to those of the Great Barrier Reef Marine Park or the De Hoop MPA but with longer corridors. They've been trying to establish a Baja to Bering Sea Marine Conservation Initiative, which would create a network of marine reserves along the west coast of North America all the way from Mexico to Alaska. Within the highly productive ecosystem of the Georgia Strait, the Georgia Strait Alliance continues to press for the creation of the Orca Pass International Stewardship Area, which would include the San Juan and southern Gulf Islands. The great predator for which the area is best known, the beautiful orca, or killer whale, is now down to only eighty-three individuals. It's clearly time to do something for their habitat.

Even the Great Barrier Reef, one of the world's most exemplary MPAS and the largest effectively protected marine park in the world, isn't able to put the ecosystem ahead of immediate human use. Trawlers, the worst monsters of the deep, that rip up the bottom and utterly destroy the marine future, are kept out of only about half of the park, up from about a quarter at the beginning! The industrial fleet had its hand out for subsidies, and AU$7.3 million of federal money had to be shelled out to restrict the trawlers even to that degree. A study found that a single pass of one of these trawl nets removed up to 25 percent of all the organisms on the ocean floor, and thirteen passes destroyed 90 percent of all its life-forms. These trawlers are often after prawns, which means they will toss six to ten tons of bycatch overboard for every ton of prawns they keep. Phil Cadwallader, Director of Fisheries at the Great Barrier Reef Marine Park asks, "Is that appropriate in a World Heritage area? Yet no politician will ban trawling in the marine park."[13]

Cadwallader says that local environmentalists were outraged when government money was used to pay industry to keep out and *not* damage a World Heritage area. He points out that the government had encouraged the trawlers to get into business in the first place by subsidizing them, so they had to help them get out. The issue is tricky, because limiting the number of trawlers just inspires the remaining ones to upgrade their technology. That's why penalty provisions are now in place to counteract the attractiveness of more powerful radar and other technological improvements. Fines set previously at AU$16,000 for illegal trawling are going up to close to a million dollars, with the potential suspension of commercial fishing licenses. There are even transponders with satellite linkups that tell enforcement officers on shore the exact location of each trawler. Cadwallader says with a small smile, "There are complaints about Big Brother... but it's not compulsory to do all this. Only if you want to fish in here, in a World Heritage area." And of course the industry does want to fish here because reserves crank out the fish. Canada has one of the longest national coastlines on Earth, but it still has virtually no MPAs that are effectively patrolled and few that are worthy of the name. There's much obfuscation on the DFO websites about "studies," "intentions," and "future preserves." The Haida have been trying to get the marine areas off the UNESCO World Heritage Site of Gwaii Haanas protected "from mountaintop to ocean floor," and the Wemindji Cree are trying hard to establish an MPA in James Bay. But Canada's federal government has only added one MPA, Lancaster Sound, and a preliminary agreement with the Haida very recently. In the twenty-some years it's had to "study" MPAs, this is a scandal.[14]

Canada isn't alone. Around the world, with the possible exception of Kenya's little preserve, parts of Florida, or Costa Rica (the country recognized as having the best consolidated system), there are very few places with exemplary fisheries and oceans policies in place upon which we might model others. Even the usual shining examples of social and environmental responsibility, Norway, Denmark, and Iceland, are actually leading efforts to murder whales and other large marine mammals, killing even more than the whalers of Japan, who slaughter thousands of animals a year for "research purposes." That's in order to get around the whaling ban of the International Whaling Commission, to which these four countries are all signatories. Where do they kill them? Inside supposed MPAs or national preserves, like a huge one extending down to the Antarctic from Australia. Australia should be patrolling this area. Instead it turns a blind eye to the Japanese as they plunder what's left of the highest trophic levels of marine life.

This blind eye afflicts most of the governments on Earth. Journalist George Monbiot had to watch scallop dredgers in summer 2009 destroy the seafloor of Cardigan Bay, off the coast of his native Wales. He wrote, "The Cardigan Bay dolphins are one of only two substantial resident populations left in British seas. It is partly for their sake that most of the coastal waters of the bay are classified as special areas of conservation (SACs). This grants them the strictest protection available under European Union law." Looks great on paper, but typical of MPA legislation, reality is something else again. Monbiot asked piteously, "The bay is strictly protected. It can't be damaged, and the dolphins and other rare marine life can't be disturbed. So why the heck has a fleet of scallop dredgers been allowed to rip it to pieces?"[15]

Monbiot described how his neighbors and a group called Friends of Cardigan Bay harassed the Countryside Council for Wales to do something; the council moaned in reply that their powers aren't explicit, that "the precautionary principle," which they're supposed to be exercising, "is a vague term," and put off a decision until after the end of the dredging season. "In twenty-four years

of journalism I have not come across a starker example of bureaucratic cowardice," Monbiot fumed. "The boats are not resident here.
They move around the coastline trashing one habitat after another.
They will fish until there is nothing left to destroy, then move to the
next functioning ecosystem. If, in a few decades, the scallops here
recover, they'll return to tear this place up again." He wrote about
the noble claims of government-sponsored MPA movements around
the world and concluded, "as I have seen in Cardigan Bay, it doesn't
matter what they say they'll do if no one is prepared to enforce it."
And with a very few exceptions, there's only one person on this
entire planet who is actually enforcing marine protection law.

SALTWATER HERO

I never realized how much he risked, including his own safety. He
literally walks into the lion's mouth like Daniel and continues to
speak truth to power and call them out. I have never seen anyone
ever challenge authority or challenge evil that way. And he pays
the price for it.

Actor and Sea Shepherd supporter Martin Sheen[16]

Readers may have seen an extremely popular television program
on the cable channel Animal Planet called *Whale Wars*. It's a simple
reality show. A film crew gets on board one of the two ships operated by the Sea Shepherd Conservation Society, a marine NGO, and
follows it around as its crew picks up people, repairs its often failing
equipment, gets lost in ice fogs or trapped by glaciers, searches for
whalers on the trackless seas, and confronts, head-on, the Japanese
whaling fleet, shooting rancid butter at the harpoon decks and physically interposing their little inflatables in the icy waters between the
harpooners and the whales. It all takes place in what is supposed
to be a no-kill preserve claimed (but not adequately protected) by

Australia, deep into the Antarctic. One episode showed returning crew members and Discovery Channel filmmakers having their entire season's film seized by the Australian government at the behest of the Japanese (they went to court and got it returned); another shows a bullet fired by the Japanese nearly killing Paul Watson, the famous founder of Sea Shepherd. It's nail-biting action, and along with a recent feature called *Sharkwater* on the slaughter of another marine species, it's making Watson a global star.

Born in Canada's Maritimes, Paul Watson has been known for thirty years as an "environmental radical," destructive of property, although strictly nonviolent and even comical when it comes to his organization's actions against people. He helped found Greenpeace, but when it turned away from direct action campaigns, it lost Watson. To this day he remains the only person who has taken his helpless outrage at our treatment of the creatures who live in our oceans to the next step: physical confrontation. Watson sunk half the Icelandic whaling fleet in the 1980s and has become the world's expert on ramming, being more practiced at it than any modern navy. He does so to damage and cripple the huge boats illegally harvesting sea life so that they have to limp back to port for repairs, thereby cutting into their short killing season. Every year Sea Shepherd's activities are estimated to save the lives of around five hundred whales, including the huge majority of the population of endangered fin whales, as well as thousands of sharks, dolphins, and other forms of sea life. Until the next hunting season, that is, when, with the connivance of the very governments that have signed whaling bans or that supposedly defend MPAs, whaling fleets manned by Norwegians, Icelanders, or Japanese fishers set off to plunder again.

Marine equipment, that the Japanese have taken to ramming, at considerable danger to his crew's lives, doesn't come cheap. Paul Watson's NGO nonetheless devotes huge amounts of energy and money to protecting the creatures that live in our oceans. He doesn't worry too much about the way media and government have demonized him, saying, "we are answerable only to our clients." He doesn't mean Sea Shepherd members, the coastal communities asking for help, or financial supporters. He means, "whales, sharks, sea turtles, seabirds, dolphins, seals, fish, invertebrates, and plankton." Watson writes that "the Sea Shepherd Conservation Society has been in existence since August 1977 and during that entire period... we have never injured a single person, sustained a single serious injury, or been convicted of a felony crime anywhere in the world. Despite this, the stories are spread throughout the media that we are pirates, extremists, criminals, and even 'terrorists.' Why is this? Because our small non-governmental organization is the only conservation organization in the world that has the guts to take on superpowers and organized crime like the Yakuza, the shark fin mafia, and corrupt politicians."

In a wash of depressing news about the state of our oceans, the Sea Shepherd Conservation Society—tiny, unique, embattled, indomitable—is one very rare piece of good news. The two ships and one helicopter that Sea Shepherd deploys operate in open zones or in ones that are supposedly under some sort of official protection. Australia can't complain when Watson chases off Japanese monster boats that are taking whales out of their waters illegally. Watson's popularity in that country and in neighboring New Zealand, where he enjoys support from the influential indigenous Maori community, has prevented the government from denying him access to their ports. There he generally receives a hero's welcome from the populace; the Australian government may not care about the whale populations, but most Aussies do.

So, although vilified as a pirate by the whalers he pursues and certainly by Canada's sealers, and often arrested when he comes into port, Watson is also a hero in the many countries that solicit his help, such as Ecuador, Chile, and Brazil, as well as Senegal, South Africa, Singapore, and Dubai. In 2000 Sea Shepherd signed an agreement with Ecuador to help that nation patrol and protect the Galapagos Islands, another World Heritage site being targeted by poachers, "eco-tourism" operators, and pirates seeking to kill sharks, the famous but endangered Galapagos tortoises, the islands' giant marine iguanas, and sea cucumbers. Sea Shepherd donated a 95-foot patrol boat and a full-time officer to the Galapagos almost a decade ago. Since then this work has extended to Ecuador's mainland, where Sea Shepherd helps the Ecuadorian Environmental Police protect one of the rarest mammals on Earth, the Amazon pink dolphin. In 2007 the country awarded Paul Watson the Amazon Peace Prize for his work on both the islands and the mainland, and now, thanks to the *Whale Wars* series, he's also become a hero in places like the U.K., the U.S., Australia, and New Zealand. But largely because of his long opposition to the Canadian seal hunt, a Mom-and-apple-pie voting issue in the Maritimes, he remains a pariah in his own country. Even the CBC is remarkably biased against Sea Shepherd's work.[17]

In winter 2007, Watson was preparing to sail out of Australia to try to intercept the killing machine of the Japanese whaling fleet as they arrived in Australian waters to illegally slaughter whales. For most of a decade, Watson had been sailing under the flag of his own country, Canada. But just as he was ready to depart, and without any warning, newly elected Prime Minister Stephen Harper pulled his boat's registration. Harper gave no legal justification for doing so, but it was so late in the game Watson had no choice but to set sail or let the whales die. Without a formal registration, he would be on the high seas illegally himself, subject to any form of violence that the Japanese, or any other enemy, would care to dish out.

A few months previously, two representatives of a small group of Iroquois traditionalists from Quebec, Stuart Myiow Junior and Senior, had attended a Sierra Club of Canada national conference to deliver a workshop on native beliefs. Watson was a keynote speaker there, and they were in the audience. The Myiows were electrified by his account of his life and work. Following the talk they approached him as he was rushing off to catch a plane to get back to sea and asked if they could perform a special protection ceremony and present him with a medicine bag. The ceremony was an unusual and moving moment out on the grounds of the college hosting the conference, with Watson's ride to the airport waiting in the background. "We wanted to help somehow," Stuart Myiow Junior explained, "and all we could think to do was one of our most powerful ceremonies of protection. I don't know why I brought the medicine bag with me, and the tobacco and all that we needed—I just did." After this encounter, Myiow's longhouse, the small Mohawk Traditional Council on the Kahnawake Reserve just outside of Montreal, continued to follow Sea Shepherd's activities on their website. When Harper pulled Watson's flag, these friends racked their brains for what to do. "We thought, well, the Iroquois are a sovereign nation, as defined by our treaties with Canada; we have the right to travel without passports and to bargain with the government as equals. So why not register a ship?" They wrote Paul Watson, offering him a flag and registration documents. The flag was sent at once, but the full registration had to wait until the following June, when Watson made a rare trip to Eastern Canada to receive his flag.

The Mohawk Traditional Council experienced their own adventures with the flag and registry. Like most grassroots groups passionate about the environment, they're a small minority in the Kahnawake community, which is also only one of the eight reserves of the six nations of the Iroquois Confederacy. Other native people both in and outside their reserve challenged their right to speak for everyone and issue a flag in the entire confederacy's name. Stuart Myiow Junior made what turned out to be the perfect reply. He told his critics they were absolutely right: he and his group didn't have that kind of power. But the nation as a whole did. While they might start the very long process of discussing it and trying to come to a consensus, the whales would be killed.

"It came down to a choice between doing what little we can, as soon as possible, or doing nothing and letting the animals die," Stuart said. And the critics, to their everlasting credit, agreed. The flag was presented in the name of the entire Iroquois Confederacy, and it flies in full sight of the *Whale Wars* cameras today, an object of pride to many members of the Canadian aboriginal community and of course to that small group that wanted so much to help. As for Watson, he says that in all his life no one has ever done anything quite like that for him before—especially someone from his own country. He's even placed the story of his alliance with the Mohawk Traditional Council on Sea Shepherd's website, right next to another treasured letter of support, from the Dalai Lama.

The oceans are huge. Each one of us feels very small in comparison. Most of us spend our lives far away from the slaughtered wildlife and acidifying reefs, on land. And most of us don't have much money or political influence. But as Paul Watson and the Mohawk Traditional Council have demonstrated, that's no excuse not to do whatever we can to help, given the trouble our oceans are in. If much larger numbers of us actually do try, there's no telling how many creatures and how much ocean we can save.

BAKING A
SUSTAINABLE PUDDING

AGRICULTURE

THE RIGHTEOUS PORK CHOP

The first Ark product from Mali is the Dogon shallot, a crop whose cultivation dates back to the pre-colonial period, when it performed important social functions... as a form of currency for trade, in propitiatory and magic rites and... as a basic ingredient in medicinal preparations.

Description of a Slow Food Ark of Taste foodstuff "at risk of extinction"[1]

The ongoing movement to reform agriculture worldwide is simply huge and is one of the most encouraging developments of the early twenty-first century. In fact it's arguably the single best reason for hope this planet has got at the moment. It's local, it's global, it's creative, and it's multifaceted. Unlike the confusing and even dangerous aspects of many energy technologies, the food revolution is getting almost every element exactly right in terms of balancing food production and consumption against the health of our water, soil, and air, even including social and economic concerns. Because all these elements are so closely intertwined, the more the new agricultural revolution grows, the better off human health and ecosystem health can become as well.

A multitude of changes in eating and growing practices over the past few years have come from new businesses, international NGOs, community organizations, urban gardens, and market and food outlets. A flood of books and films has led the way. Authors Eric Schlosser and John Banzhaf popularized the work of earlier researchers such as Vandana Shiva, Pat Mooney, and Ronnie Cummins by attacking the negative aspects of current food production systems. Books like *Fast Food Nation, Chew on This: Everything You Don't Want to Know about Fast Food,* and *Spoiled: the Dangerous Truth about a Food Chain Gone Wild,* as well as films like *Food, Inc.* and *Super Size Me,* have highlighted everything from the hidden calories and carcinogens in burgers and lattes to the food industry's vast dependence on corn sweeteners and trans fats. Such works have demonstrated how fast-food ingredients have led to obesity and an epidemic of heart disease and diabetes. They've demonstrated how the use of centralized slaughterhouses and vegetable-packing factories can spread contaminants like *Salmonella* and *Listeria.*

These popular books and films, read and seen by millions, have also exposed the effects of "agrifood" industrialization on land, air, water, and sea—the ocean dead zones caused by pesticides and herbicides, the contamination of soil by industrial fertilizers, and the misuse and fouling of water everywhere. They have alerted us to the fact that thousands of locally adapted or multiuse species, from Dexter cattle and Araucana chickens to ancient grains and fruit varieties, have disappeared from our farms and our diets. These breeds and varieties produced delicious food, but the grains didn't grow fast enough for impatient investors and marketing boards, the wonderful fruits were too fragile to travel thousands of miles in fossil fuel–powered trucks, and the animals couldn't survive the grotesque crowding, minimal attention, and chemical feeds of an industrial farm.

Growing food without the addition of petrochemicals remains at the heart and soul of the entire organic food movement. But organic certification, difficult and laudable as it is, sometimes doesn't address transport, processing, or cruelty issues. The U.S. FDA, for example, keeps trying to force organic growers to accept genetically engineered or municipal sludge–fertilized crops as organic; and producers with huge, crowded barns, where dairy cows spend their entire lives chained in stalls, only have to provide their captives with feed grown without pesticides or fertilizers to get organic standing in some countries. This is how low standards in a few localities, confusing claims, and huge travel distances—as when "organic" greens grown in China are mass-imported by retailers like Walmart—gave birth to the locavore movement.

Authors Michael Pollan and Barbara Kingsolver became household names with *The Omnivore's Dilemma* and *Animal, Vegetable, Miracle: A Year of Food Life*. The title of Alisa Smith and J.B. Mackinnon's book, *The 100-Mile Diet* (called *Plenty* in the U.S.), has become the term for a lifestyle exercise that millions of people are trying out. This movement to eat seasonally and only what your locality can produce has led to a great deal of media attention, including a book with the bodacious title *The Righteous Porkchop: Finding a Life and Good Food beyond Factory Farms*. Attempting to eat locally is a difficult discipline (especially when it comes to our tropical addictions, like coffee and sugar), but it does appear to be the best way to save soil, water, plants, animals, and our own health. Supporting local farmers, buying organic, cruelty-free, and fair-trade varieties whenever possible—but always from local sources—has become a new form of ethical living.

The chief way this can be done, given the difficulty of identifying the provenance of foods in most supermarkets, is to get involved directly with your own food supply by supporting farmers' markets and especially by committing to a particular farmer and prepaying for weekly supplies of produce, fruit, or meat and eggs every year. Community-supported agriculture (CSA) ventures began in Japan in the 1980s, partly because the country's lightly cooked cuisine demands such varied and ultrafresh ingredients. It remains the greatest Japanese gift to the environmental movement so far. The speed with which this decentralized, leaderless, unfunded system caught fire is just about unprecedented. Today there is hardly a person in the industrialized world, even in remote and small cities, who hasn't heard of CSAs, and there are many millions of subscribers, each with his or her own "family farmer."

Getting paid in advance for a summer's supply of produce or a winter's supply of meat and eggs gives the farmer some security in the face of droughts, onslaughts of pests, cold snaps, or flooding. If the spinach crop dries up or the chicken runs into an owl, the farmer is allowed to put more zucchini and beef in the buyer's basket instead of greens and eggs. These days, however, the farmer's major economic enemy is the increasingly integrated intermediaries who take a huge bite for marketing, transporting, packaging, slaughtering, storing, or processing crops and livestock. Every conventional farmer growing supermarket food is extremely vulnerable to changes in huge, globalized markets and to unfair consolidation in the supply chain by intermediaries, who frequently buy up all the slaughterhouses or packaging facilities in a wide area so that they can charge farmers whatever they like for their services. Retailers usually take the biggest bite of all. Moreover, perverse "hygiene" regulations often make small-scale, on-farm selling or packaging illegal, thus penalizing small producers and favoring the processing giants. Good food provided by most farmers can be contaminated in centralized slaughtering and processing facilities, which may serve most of a continent. That's the major reason more country-wide outbreaks of *E. coli*, *Salmonella*, and *Listeria* have occurred since the food system became so heavily industrialized.

All of these problems are virtually eliminated by buying locally through farmers' markets and CSAS, which are also closer to what Europeans, with their local, fresh markets, have always done. "You've begun to attack dependency and distancing in the food supply," says Brewster Kneen, longtime food activist and author of another great title, *Farmaggedon*. "You eliminate the debt factor," because farmers don't have to borrow from banks or governments to get their seed and supplies in the spring. They just have to be sure to manage their money so they're left with a profit.[2] This setup is vastly more sustainable for them because it's more fluid and flexible; city people learn that there are a lot more food varieties out there than they thought, that nature is unpredictable, and that food producers have to face challenges. Farmers learn what people really want and can quickly adjust their production to serve their market. Many CSAS are cooperatives, where members can drive out to the farm, help with weed and insect control, or just picnic in the country; most have festivals or potlucks at the end of the season and are community oriented, donating their surpluses to poorer clients or charity. "What you're really doing," says Kneen, "is reconstituting a community and a social life around food. You're creating a whole new culture. We have to realize that's what we're doing; and that's what we need."

Another consumer revolution is exemplified by the Slow Food movement, which began in northern Italy about twenty years ago. At first glance it seems less overtly an alternative to existing marketing and production systems and more like a pleasure- and history-centered club for well-heeled epicureans. Its chapters in 132 countries are called "Convivia," from the word "convivial." Their primary function is to bring people together for jolly dinners based on delicious local foods, usually served with plenty of wine. But there is substance behind the movement as well.

Slow Food's slogan, "Good, Clean, Fair," seems no more threatening than Mom and apple pie, and its "Ark of Taste" is an ongoing collection of food products from eco-regions around the world that celebrates local culinary traditions and foods while also attempting to include them in a "virtual ark," to help preserve them. Besides the Dogon onion from Mali described above, the Ark program features delights such as Canada's own saskatoon berry and Brazil's umbu and pequi fruits. However delicious, intriguing, or even effete these foods may sound, a reader might wonder whether too demanding a market might outstrip the supply of these new delicacies; but the Slow Food organization already considered all that. In fact, it has put more thought than one would expect into investigating how food is raised and how programs can be established to support this diversity. An entry on the Slow Food website honoring its "Fifth World Ham Congress," for example, describes as its goal the investigation of "the challenges that pig farming and ham production... have been facing in recent years... [such as] the deforestation of ancestral oak forests that once fed free-range pigs in Andalusia, the appearance of the H1N1 swine flu and the... need to take drastic measures... concerning the current agrifood production system." It even pledges to support "the creation of a network of farmers raising native breeds throughout Europe, to defend and promote sustainable pig farming."[3]

The Slow Food movement, CSAs, and health food co-ops, along with the popular media they work with, are all responsible for alerting the public to the dubiousness of processed foods, fast foods, and non-certified cocoa and coffee. After years of pressure, organic and fair-trade coffees, teas, cocoa, and sugar are now readily available—and not just through health food stores, co-ops, and specialty shops selling "green" brands like Divine Chocolate and Equal Exchange. Mainstream retail giants have joined the revolution. Starbucks was singled out during demonstrations in Seattle in 2000 for its resistance to fair-trade certified products. It claimed it was paying a fair price to "its own" farmers, but with no outside policing, that reassurance sounded pretty hollow to food activists. Today the massive coffee chain is offering several fair-trade brands with the appropriate certification, and is moving toward providing more.

Other giant retailers, including McDonald's, Dunkin' Donuts, and Cadbury, are introducing fair-trade products. In March 2009 Cadbury Dairy Milk announced it would begin buying its cocoa from fair-trade producers in Ghana. This part of the company sells more than 300 million bars of chocolate annually in Great Britain, which means that nearly 20 percent of Ghana's 700,000 small cocoa farmers will benefit from this gigantic new market. The very next month, Mars, Inc. announced new sustainability commitments for cocoa, moving beyond fair-trade certification to Rainforest Alliance certification, which supervises growing as well as labor standards. There is only one Mars branch scheduled to use the new cocoa, Galaxy Chocolate in the U.K., but this crack in a once stubbornly closed wall of indifference is already leading to calls from the International Labor Rights Forum and organic activists to vastly increase the number of such products, and to bring in still-resistant competitors like Hershey.

The effect these changes in corporate behavior will have on the harsh lives of small coffee and cocoa growers is hard to overestimate. Cadbury is calling its move a "virtuous circle" of purchase;

they've put a floor on prices so that they can't drop below production costs, and if a producer improves quality, it's directly rewarded with better prices. Before, such efforts were ignored by intermediaries and not worth the extra effort for growers. More use of certified products also opens the way to far more sustainable farming practices. Fair-trade certifiers will help producers manage plantations in the way coffee and cocoa evolved, as part of a complex of trees and other tropical forest plants that provide homes to birds and wildlife, a form of traditional agroforestry. They'll provide more sustainable jobs, protect habitat, and still make profits—while producing dark, delicious chocolate, which is, let's face it, one of the greatest joys of food life.[4]

SAVE THE FOOD!

In 2009 15 new Presidia were created worldwide, bringing the total number to 314. The first Presidium in Kenya was established to defend a local poultry breed... in Tajikistan, a Presidium has been established in the Pamir Mountains to protect more than 60 varieties of mulberry... Considerable work was also done in northern Europe, with new Presidia started in Austria, Netherlands, Germany and Switzerland.

Slow Food and Terre Madre newsletter, December 2009[5]

Slow Food began as a protest against its opposite, fast food, as exemplified by a branch of McDonald's that opened at the base of Rome's Spanish Steps in the 1980s. Slow Food's founder Carlo Petrini didn't stage demonstrations on that occasion; he decided to take a more subversive tack. Like Michael Pollan, Eric Schlosser, and Alice Waters, who are all active members of the Slow Food movement, Petrini decided to show how much more delightful life could be *without* the dangerous chemicals and greedy intermediaries of our current food system. On its list of organizational objectives, following the succulent descriptions of Convivias, exotic foods, festivals, celebrations, and "promoting 'taste education,'" Slow Food describes an ambitious agenda that includes "educating consumers [and] citizens about the risks of fast food... the drawbacks of commercial agribusiness and factory farms... [and] the risks of monoculture and reliance on too few genomes or varieties; developing various political programs to preserve family farms; lobbying for the inclusion of organic farming concerns within agricultural policy; and lobbying against the use of pesticides... [and] against government funding of genetic engineering." Sounds more like Greenpeace, the Institute for Agriculture and Trade Policy, or the Third World Network—but wearing that pleasant mask of conviviality.

One of the movement's initiatives is Presidia, a food standards labeling program. The requirements to use the Presidia label are long and far more complex than those for becoming certified organic. The producers join what is called a "Presidium" and "comply with production rules that respect traditions and environmental sustainability." They must, of course, use "organic or integrated cultivation methods," but they must also raise only native varieties and breeds, use raw milk in their cheeses, and pasture sheep and goats on mountainsides "to maintain regional integrity and defend the mountain ecosystem." The hay, forage, and other feedstuffs must be natural; "the use of silage or genetically modified products is not permitted."

Presidia members may not use additives, sweeteners, emulsifiers, stabilizers, or colorants, and they must use "recycled paper, glass, wood and any other material that is recyclable" for their packaging, labels, and brochures. Even fishers must follow ecological cycles and must "avoid fishing in periods when the survival of species might be endangered and only use traditional nets and fishing methods of small-scale coastal fishing." Finally, Presidia producers must respect social customs and history; each one must be connected "to a specific geographical area" and must honor its "environmental, historical, and socioeconomic perspective." In order to accomplish this, the Slow Food Foundation for Biodiversity provides veterinarians, agronomists, and food technologists for "assistance and advice."[6]

Slow Food finances itself by producing popular consumer guides and publications for sale. Slow Food's *Gambero Rosso* (Red Crab) guides to wine and restaurants have become to Italy what Michelin guides are to France. They're gobbled up every month and can make or break a vintage or a business. Its many food salons such as the *Salone del Gusto* in Turin and its yearly international *Terra Madre* festival attract hundreds of food and wine producers each year, each happy to pay a fee to show off their wares. In such simple ways this movement has become so powerful that when the EU tried to enforce rigid "hygiene" standards tailored to favor large processors and producers, Slow Food was able to fight back.

People in North America are unhappily familiar with these "scientifically based" hygiene standards, which were invented by NASA, the U.S. space agency, in a particularly unnatural context: to keep astronauts from getting sick. They were then adopted in other unnatural environments, like corporate food processing factories at Kraft and Unilever. As a recent health profession article put it, introducing such requirements in the EU would have imposed "impossible burdens of reporting, paperwork, and new equipment... driving thousands of small farmers out of business." This would not really have been an unintended consequence, since "health protection" is often the cover that accompanies centralization and industrialization schemes that make life impossible for small farmers and very handy for food giants. It didn't work in this case, at least partly because Europeans suffer and die far less often from food-borne illnesses than North Americans. In the U.S., the origin of the NASA standards, more than 25 out of 100 people are struck by food poisoning every year, as opposed to France's food poisoning rate of 1.2 per 100—almost 25 times lower.[7]

So which products prove they fit into the natural ecosystems of this planet by their wholesome effects? The irradiated, plastic-wrapped industrial agrifood flowing out of hygienic modern factories in the U.S.? Or the individual, artisan-produced products in France, where you can find skinned rabbits hanging in the sunshine at outdoor markets, mounds of just-picked carrots with dirt still on them, and mold-encrusted, air-dried sausages? In Europe, half a million healthy consumers were happy to sign a petition against the dangerous NASA rules. In North America so far, no single agency has been strong enough to oppose them, and that explains why hundreds of thousands of artisan food makers in Europe continue to prosper, while those across North America write books like Joel Salatin's *Everything I Want to Do Is Illegal.*

In an effort to support Salatin and other organic producers concerned about food quality and sustainability, venture capitalist Woody Tasch founded a nonprofit in line with Slow Food values. Today "Slow Money" invests in various initiatives because "the vast majority of sustainable-agriculture enterprises [have] little or no access to investment capital."[8] The inspiration to do this came partly from Herman Daly, a former World Bank economist who taught at the University of Maryland. There, Daly developed "Ecological Economics" as a way to place human financial activity *within* the envelope of the world's natural systems, instead of making the realities of natural law a tiny adjunct to human obsessions about finance. Currently, national economies speculate on food as a commodity and dismiss soil fertility, water quality, nutritional value, and social traditions as "externalities" to their central concerns with making money. When Daly discusses this point, he likes to describe a favorite cartoon: "A fisherman is standing with a puzzled look on his face. On the one side, the fish in the sea and in his net are saying, 'Slow down! We can't grow that fast!' and on the other side, a banker is saying, 'Speed up! We have to grow faster!' The fisherman is caught between the rate of growth of money and the natural, biological growth rates of species."

The lofty goals of Slow Money include building "a nurture capital industry," where companies have to live up to strict requirements when they apply for the legal charters that give them the right to exist. They must, for example, give away 50 percent of their profits to sustainable practices and invest 50 percent of their assets within 50 miles of where they live. In an era of bailouts for failed banks and industries, Slow Money points out that, "If it is prudent to invest tens of billions of dollars a year in a few thousand high-tech companies... then mustn't it be prudent to invest a few billion dollars a year in tens of thousands of small food enterprises that are essential to the long-term sustainability and health of soil and economy?"[9]

GETTING IT ALL EXACTLY RIGHT

Growing Power is probably the leading urban agricultural project in the United States. [It's] not just talking about what needs to be changed; it's accomplishing it.

Jerry Kaufman, professor emeritus, University of Wisconsin–Madison[10]

There are now so many exemplary revolutionary and sustainable food systems all over the world that it's hard to pick just one as a model, especially since most of the time they have little connection to or knowledge of each other. The example we've chosen below is particularly telling, because it serves a poor community in an urban neighborhood—a fact that immediately addresses some of the economic arguments that might occur to some readers about the "expensive" and "pretentious" nature of organic foods.

Growing Power is a food-producing dynamo within the city limits of Milwaukee, Wisconsin. It doesn't have the support of centuries of tradition, like Presidia cheese producers in Italy do, and it can't sell to tourists or wealthy neighbors because it doesn't have either. It isn't linked to Slow Food or any well-known food NGO. It is typical, however, in that it doesn't take up much space and it packs a wallop. Growing Power sustains a huge social system that in itself would be worth the effort, even if it didn't produce goat stew with greens, grass-fed beef burgers, and homemade bread, dripping with local honey, along the way.

Growing Power operates from fourteen greenhouses that cover only 2 acres. Besides salad greens, tomatoes, and other produce, the greenhouses shelter goats, ducks, bees, turkeys, and even fish—tilapia and Great Lakes perch, grown in an incredibly well-thought-out aquaponics system that uses the planet's precious water supply much the way Mother Nature would. A 40-acre farm in the nearby countryside grows more vegetables, as well as the hays, grasses, and legumes required to feed the livestock. Growing Power gets its soil by composting more than 6 million pounds of food waste a year, including material collected from local breweries and coffee shops. In fact, the red wiggler earthworms used to turn this compost into what the workers call "Milwaukee Black Gold" are considered part of the project's livestock.

Besides the thousands it serves in Milwaukee, Growing Power has helped set up five similar projects in impoverished areas of Arkansas, Massachusetts, and Mississippi. Its largest effort is a sister program begun in the infamous, crime-ridden Chicago housing project Cabrini-Green, which now includes a center in Chicago's downtown Grant Park, teaching young people to grow 150 varieties of heirloom vegetables, herbs, and edible flowers. All these projects are supported by selling the food in Farm-to-City Market Baskets, which is another example of that famous CSA food distribution system that links urban dwellers with their food supply.

Food is a crucial social issue for human beings because it remains the primary way that people meet, socialize, celebrate, and even grieve. And of course it's also the primary way we survive and maintain our health. Along with indicators like child mortality rates and average income, access to *good-quality* food is one of the most accurate ways to judge a country's prosperity, as well as its level of social equality and integration. In marginal, minority, or impoverished communities around the world—and this very much includes native reserves in Canada and urban ghettos in the U.S.— there is nearly always less food of good quality available.

Will Allen, the founder of Growing Power, is a recipient of a MacArthur Foundation genius award; he's a 6-foot, 7-inch ex-pro basketball star, today in his early sixties. Allen grew up as part of a poor sharecropper's family in South Carolina. Eventually they bought a farm in Maryland and proceeded to feed the neighborhood with it. "We could handle thirty people for dinner," he remembers. "Food racism" is one of the major reasons Allen created Growing Power, which provides thirty-five jobs to a very ethnically diverse staff and is the only place for many miles in any direction that carries fresh produce, free-range eggs, grass-fed beef, and homegrown honey. "Poor people are not educated about nutrition and don't have access to stores that sell nutritious food; they wind up with diabetes and heart disease," he says, and adds, "Just as there is redlining in lending, there is redlining by grocery stores, denying access to [the poor] by staying out of minority communities." Every week Growing

Power delivers around three hundred Market Baskets of food to more than twenty agencies, community centers, and pick-up sites around the city, introducing sustainable eating habits to thousands of people, getting them off expensive and poisonous industrial foods, at a price that works out to between $9 and $16 a basket.

Because his clientele likes fish, and because yellow perch are a traditional, local food that has almost disappeared owing to pollution and overfishing in the Great Lakes, Allen decided to farm fish. This is a tricky proposition in most cases—fish raised in ocean cages have been a disaster, breeding diseases and parasites that have all but wiped out their wild relatives. Moreover, their flesh is contaminated with pesticides and chemicals from their feed. Even when vegetarian species like tilapia are raised on land, what the fish eat and what happens to their wastes are often mismanaged. But Allen didn't get a genius award for nothing. He built a farming system worth $50,000 for just $3,000 inside his greenhouses. The perch hatchlings are raised up for market over about nine months in 10,000-gallon tanks filled with water that's recycled throughout the system. The fish waste drops down to a gravel bed where it creates nitrogen that feeds watercress growing in channels down the line. That filters the water some more; it's then pumped overhead to nourish tomatoes and salad greens. "The plants extract the nutrients, while the worms in the soil consume bacteria from the water, which emerges virtually pristine and flows back into the fish tanks." This multiuse growing system mimics nature through the crucial process of using waste as food and using water as both a nutrient and a waste-disposal mechanism. It also adopts traditional farming methods used around the world, maximizing a wide variety of production in a very small space.

Will Allen is part of the Growing Food and Justice for All Initiative, a network of about five hundred people working on sustainability and food access issues. He supplements Growing Power's offerings with products from the Rainbow Farmer's Cooperative, which consists of three hundred multiethnic family farms in the U.S. Midwest and south. These partners' longer seasons allow Growing Power to provide fruits and vegetables all year round. In all its centers Growing Power concentrates on training: about three thousand young people from around the world learn how to heat greenhouses with compost; use worms to create fertilizer; build fish-farming systems and hoop houses (low-cost, movable chicken pens); and exploit other low-tech, high-yield techniques that are features of all the new sustainable food-raising systems. In educational programs in Chicago and Milwaukee, Growing Power gets people of all ages involved in creating food wherever they live, in inner-city lots, suburban backyards, and rural areas. "For kids to make their own soil, grow their own food, and then get to eat it, that's a very powerful experience," says Allen proudly.

Growing Power is a particularly fine example, but there are literally millions now, in every city and in most large urban areas in the industrialized world, multiplying at a bewildering rate since the first edition of this book was written. They may not all have an aquaponics system, but most have evolved some unique means of producing food that is their own creative response to local ecosystem products, as well as to local needs. All people who care about food can name a farm, a market, or a co-op they think is wonderful in their own area. These businesses demonstrate that the regreening of the entire world of food is well underway.

IDEAS ABOUT AGRICULTURE: THE ZOMBIES

In 1965, wheat yields were 4.6 million tons in Pakistan and 12.3 million in India. By 1970, after the introduction of our new wheat, Pakistan produced nearly twice its amount, while India increased its yield to 20 million tons. The trend continues. This year Pakistan harvested 21 million tons, and India 73.5 million— all-time records.

Norman Borlaug, "Father of the Green Revolution,"
in a *Wall Street Journal* editorial[11]

The so-called Green Revolution, the massive, post–World War II transformation of agricultural practices that uses chemical fertilizers, pesticides, and monocultured crops, began to claim yields like those described above shortly after it started. But they give a very distorted picture of what's really been happening in food production since the 1950s. Growing larger "desirable" parts of a plant through selective breeding and genetic engineering has resulted in more harvestable seed, as when hybrids of wheat, say, are bred to produce huge, tasseled heads and little stems. This kind of plant does mean more wheat seed per acre, but it also means less straw and fodder, which still have to be grown somewhere. That loss to the farmer isn't tabulated. Moreover, these new varieties *always* demand more chemical fertilizer: nitrogen, phosphorus, and potassium made from fossil fuels, as well as a great deal more water. When those crops fail to perform, their creators always blame lack of water or insufficient application of fertilizers and pesticides.

People who turn to the first Green Revolution's latest technological fixes like genetic modification, or huge irrigation projects, claiming they will feed the planet's growing populations, *never* subtract the cost of lost crops, depleted watersheds, and heavy chemical inputs from their projected yields. They also routinely overestimate what technology can do. A recent overview of genetically modified (GM) crops correctly points out that "Hype and promises... notwithstanding, there is not a single commercial GM crop with increased yield, drought-tolerance, salt-tolerance,

enhanced nutrition, or other attractive... traits. Disease-resistant
GM crops are practically non-existent."[12] By contrast, modern conventional breeding techniques are coming up with drought-tolerant
cassava, a traditional rice high in iron and zinc, and higher-yielding
soya. These crops are developed by publicly funded organizations
rather than for-profit chemical conglomerates, which probably
explains their focus.[13]

Using high-tech means such as genetic engineering to deliver
more food is rapidly becoming what medical researchers call a
"zombie argument"—which is to say, it's dead. For years independent
researchers across the globe have been finding massively against
these high-tech solutions because of their downsides, like destroyed
soils and water tables and the impossibly escalating costs of their
fossil-based inputs. But the idea that genetic engineering or intensive industrial agriculture will "feed the poor" is still dug up from
its grave over and over again, for the simple reason that it serves
vested business interests. Dressed in new scarves and gloves, the
zombie is paraded before the public, who are made to feel guilty if
they withhold the blessings of the chemical industries' most ghastly
mistakes from an impoverished Africa or India.

Truly increased yields are central to the industry's argument
about its ability to create more food; and yet, all that genetic engineering has produced so far is "yield drag"—a measurable lessening
of production of between 5 and 10 percent, depending on the variety.[14] Often such losses in yields are presented as acceptable, because
the farmer saves time and costs in weed or pest control, since the
most common genetically engineered crops resist a patented her-
bicide or contain their own pest killers. However, the time-saving
factor doesn't support the claims for potential growth in the food
supply, even if increased yields per acre *were* the only way to help
more people avoid starvation.

If we were to accept that the way to feed the hungry is to increase food production by increasing yields per acre, something that's debatable, doing so by gene manipulation would require augmenting more soils with more nitrogen, the key fertilizer for plant growth. Hybrids and genetically engineered varieties of plants demand a great deal more nitrogen to function than do normal varieties. They also need much more water than conventional varieties, and that fact presents its own problems. Plant scientist and respected specialist on agricultural production E. Ann Clark, of the University of Guelph, points out that there is "an inherent conflict between our efforts to increase yields and safeguarding the environment." She explains that, "If you're using a lot of nitrogen to boost yield, you're increasing the generation of methane in the livestock that then eat such rich foods. Food crop soils [of any kind] can also leak nitrogen oxide." For example, "when the amount of nitrogen you apply exceeds what the plants can use, denitrification losses result—oxydized nitrogen, a powerful Greenhouse Gas, then escapes into the atmosphere." In short, once again, we run into biological limits. Producers can have increased yields, but one way or another, after certain natural limits are reached, people can harvest more food per acre only at the cost of our atmosphere or our water supply.[15]

If enriching fields with nitrogen to increase food yields leads to pollution and climate change, what else can we do? A new body appointed by the UN has some of the answers. The International Assessment of Agricultural Knowledge, Science and Technology for Development (IAASTD)—analogous to the Intergovernmental Panel on Climate Change (IPCC), the fount of global knowledge on climate change—is composed of four hundred international agricultural researchers. Like the IPCC it receives its funds from the World Bank, the UN, and the Global Environment Facility. Before IAASTD's hotly anticipated first study was due, the mainstream press was excitedly predicting it would champion all kinds of new techno-solutions.

Partners like the OECD and the World Bank expected the report to push for a "New Green Revolution" based on more high-tech engineering and more chemical inputs; the fertilizer and biotech industry confidently awaited a boom in business. These groups have blamed the twenty-first century's rapid downturn in supplies of food on a "decline in investment in agricultural research and infrastructure." Rather than acknowledge the possibility that the global food crisis might be the result of breaching natural system limits, the industrial and scientific mainstream—even the general population, if polls are anything to go by—have been looking to research rather than restraint for answers.

When the IAASTD report came out in April 2008, however, it did not push for more research and technology. Instead, it recognized natural system limits and suggested real, sustainable solutions, beginning with—of all things—an expansion of small-scale, organic agriculture. The IAASTD remains the most prestigious international agricultural research body in the world, and anyone interested in the global food supply should read its report. The section dealing with national food security starts by listing the reasons why world food prices in the early twenty-first century spiraled out of control; in other words, why the IAASTD's work had become so vital. Its analysis is that the crisis was fundamentally caused by the same small-government, big-business policies that have broken the financial system: trade globalization and financial and corporate deregulation policies introduced by Ronald Reagan, Brian Mulroney, and Margaret Thatcher twenty years ago.

The report lists the many forms of agricultural activities around the world that are unsustainable because they depend on heavy chemical inputs, export markets, and irrigation with imported water. Also typical of these unsustainable forms of agriculture are landgrabs and international speculation on food. The IAASTD's fundamental criticism of the current paradigm is that "international trade in agricultural commodities as currently organized [for example, the big boom in investing in biofuels] sets consumers in different countries into competition for the same land and water resources." China and the Arab Gulf states are buying up agricultural land all over the world to feed (or fuel) their own populations. The question is, what land will the countries they're buying from, in Peru, Southeast Asia, Africa, and even the U.S., now use to feed their own populations?

From Turkey and Pakistan through Thailand, and the Philippines, government elites in desperately poor regions are giving away the land beneath their feet to anyone who will pay them to lease or buy it. In June 2009, when Arab investors lost money during the economic downturn, the Turkish minister of agriculture and rural affairs, Mehdi Eker, declared open house. "We have made maps of all our lands. Take and cultivate which you want."[17] The wealthy Gulf oil states and China are leading the landgrab, but U.S. billionaires like Bill Gates aren't far behind, using a supposedly "mutually beneficial" organization called AGRA (the Alliance for a Green Revolution in Africa) to acquire the best farmland on the continent; although one can imagine that chemical companies will benefit most from "developing Africa's breadbasket" with Green Revolution technologies.

The situation in Africa has gotten so serious that many countries have placed *de facto* moratoriums on land purchases until they can better assess the situation; but 6.2 million acres of farmland in five sub-Saharan African countries have already gone missing. The UN's Special Rapporteur on the Right to Food has called for a new set of human rights principles to control landgrabbing. Canada has its own reasons to join the Third World in supporting building strong international oversight. In fall 2009 Chinese investors were roaming the farmlands south of Montreal, trying to buy up big chunks of what little agricultural land Quebec has been able to preserve, putting its *Loi sur la protection du territoire agricole* to a severe test.[18]

The IAASTD report suggests the best way to reverse this trend is to support collectives of small farmers by "improv[ing] *tenure* and access to resources." In short, by making sure local people, not distant multinationals or richer countries, own and manage their local food supplies. They ask that the international community "address market concentrations, especially in grain markets, at the global level," pointing out that "Investment in the agricultural sector has focused largely on export crops to generate foreign exchange, forcing countries to rely on continued low international food prices to meet national food demand. That strategy has failed."[19] The last statement is incredibly strong, considering the IAASTD's exalted position and the expectations of the actors that funded it. It's strong because the four hundred researchers had to base their recommendations on global reality, not financial pipe dreams.

Today six private multinational corporations—Bayer, Syngenta, BASF, Dow, Monsanto, and DuPont—control over 80 percent of the agrochemicals produced in the world. In 1994 there were thirteen agro-giants, but mergers and consolidation in what is aptly termed the "agrochemical industry" reduced these and continue to reduce the field today. Such a concentration of ownership places opportunities for price fixing and major food policy decisions in the hands of profit-obsessed corporations rather than governments or farmers. Moreover, power over *a third* of all the staple grain seed produced globally is shared by only four firms, three of which, Monsanto,

DuPont, and Syngenta, also control the fertilizers and chemicals. Is it any wonder that these corporations have only developed seeds that need chemicals and have concentrated on seeds and inputs that they can patent and withhold from producers, depending upon their profit needs?

Potential good news abounds in the IAASTD's suggestion that "market concentrations" must be addressed—a mainstream acknowledgment that profits, not food quality or need, are the only concern of the biggest food players.[20] In support of the IAASTD's concern with actual food production, the report even suggests that governments "Mobilize the capacities of supermarkets and other public and private actors along value-adding chains to offer consumers affordable, safe, healthy, fair-trade foodstuffs that demonstrate commitment to poverty reduction, environmental and climate change goals." That makes it clear that this UN body favors a Growing Power, rather than a Maple Leaf, Smithfield, or Walmart business model.

But the enormous power over life's necessities, food and water, so recently gained by corporate giants, is not going to be relinquished easily. As in many other aspects of the battle for the planet's survival, the way that the wrong management paradigm is organized—centralized, chemically based, privately controlled, fossil fuel–dependent—is going to respond less to a direct assault and more to the same kind of gradual and subtle changes that saved Europe from the U.S. model of hygiene control. This subtle approach is also part of the IAASTD report, which has a separate section devoted exclusively to something that sounds harmless and bland: "*multifunctional* agriculture for social, environmental and economic sustainability."[21] That simple term, "multifunctional agriculture," is positively Earth shaking in the new paradigm of food production it portends.

IDEAS ABOUT AGRICULTURE: THE LIVE ONES

The concept of multifunctionality recognizes agriculture as a multi-output activity producing not only commodities (food, feed, fibers, agrofuels, medicinal products and ornamentals) but also non-commodity outputs such as environmental services, landscape amenities, and cultural heritages.

IAASTD report Executive Summary, April 2008[22]

Imagine terraced orchards growing lemons in the Italian sunshine. Nearby are grapevines twining on wrought-iron arbors, with a few hairy, dark red pigs snuffling behind a wattle fence near the end of one terrace. It borders a gorge crossed by a narrow stone bridge with a road leading to a medieval church. It's dusk, so the air is alive with swallows; hundreds of bats are starting to wheel out of the church belfry to the sound of bells coming from the village in the gorge below. Or imagine rice paddies in Bali, brimming with small fish, home to herons and egrets, dotted at sunset by children collecting snails or netting fish for supper, while in the still shining hills above, a procession of costumed priests and dancers, clanging and jingling, winds past the local temple to the water goddess. Or imagine something closer to home: a village in New England or southwestern Quebec, with a white-steepled or stone church. It's late afternoon in the fall, so children are pouring out of the local school and boarding buses that will take them home through a covered bridge, past apple orchards, stone fences, and the spooky old county graveyard.

These are the "services, landscape amenities, and cultural heritages" the IAASTD report is talking about. It is local farmers, not vertically integrated corporations or their employees, who lovingly maintain churches and vineyards. To do so, they have to be able to support themselves in the countryside, and that generally means having ownership of farmland. The larger the farms, the fewer people there are left to plant the arbors, feed the pigs, keep the streambeds clean, and take animals out to pasture. Because "agricultural multifunctionality" enshrines these activities and requires reforming international farm policy, it has alarmed the three countries most committed to the spread of industrialized, highly unsustainable agriculture—the kind that uses full vertical integration, corporate ownership, heavy machinery, petrochemical poisons, long-distance transport, genetic engineering, and monocultures. These countries—Canada, the U.S., and Australia—have become known as the hated spoilers, not just of climate change agreements but of every United Nations attempt to protect biodiversity or to deflect agriculture from its current suicidal path. Unlike the other 107 countries involved, they refused to approve the IAASTD report.

That's because multifunctionality, by definition, would discourage vertical integration and would favor the small farmer, local slaughterhouses and storage facilities, diversity of production, and natural biodiversity. Large corporations, with the collusion of our governments, have worked hard for the past sixty years to eliminate people from the country landscape and replace them with machines and monoculture management. But this report by four hundred international experts recognizes that *only* farmers running small or medium-sized operations can provide the personal attention that a multifaceted and multiproductive approach to agriculture demands so that the land base remains both economically and environmentally sustainable over the long term. The industrialization policies of the old Green Revolution were consciously intended to get cheaper food out of the countryside and use the displaced farmers to provide new industries with a bigger labor force. But today these policies are working to the tragic disadvantage of the citizens whose countries are still applying them. Most Third World cities are swelling so fast with refugees from the seizure of the countryside by big operations that they are topping 12 and 15 million people, a population density widely recognized to be both unmanageable and unsustainable. And good jobs are not available to the huge majority of these refugees.

By contrast, the multifunctional approach is based on scientific studies and hard policies that were adopted by consensus in the EU over a decade ago. Multifunctionality has made European farmers the most financially secure in the world, their landscapes among the most beautiful and desirable, their countrysides the most livable, and their food of the highest quality. Critics of multifunctionality claim that the concept is protectionism in disguise. They have turned to trade regimes like NAFTA and the WTO to prevent its spread, although even the WTO has rarely found such practitioners in noncompliance.

A major argument waiting in the wings throughout this discussion is encapsulated in a post to an NGO website: "Isn't it a bit dubious to hear wealthy people living in monocultural landscapes, the destruction of which made them wealthy, dictate to the world's poor on how to value ecosystem services, landscapes, and cultures— things for which there is no market?"[23] It might be possible to argue in favor of the spread of monocultures if all the world's poor people really could become rich (or even fed) by thus despoiling their home landscapes. But there are a variety of reasons why they can't, all related to the crises in usable soil and water, the unavoidable future escalation in the cost of fossil fuel supplies, and the simple fact that large monocultures favor large landowners. Even if industrial agriculture miraculously survives the steady increase in the supply of fuel and chemical inputs, policies favoring it are far more likely to push poor people off their ancestral lands and into urban slums than to make them rich. Thanks to unionization in nineteenth- and twentieth-century Europe and North America, urban employment did generate wealth for a generation or two. However, moving from the land to the city to get a good hourly wage in a factory (or call center) would be seen as poor career planning for most refugees from rural areas. There aren't enough jobs to go around, and since the downturn, many such refugees, in China, for example, have turned around and headed for home.

Leaving aside the heated discussions about rich versus poor and city versus country, the problem of how to provide everyone on the planet with a Western standard of living is one that is amenable to mathematics. Given what each First World citizen consumes, the number of people who are now alive, and how many natural resources remain for us all to divvy up, it would take (according to variations in calculations) between at least four and as many as *eight more planets* fully as rich in natural resources as this one. Right this minute, even with many countries not suffering from a food deficit, we would need the current planet's total production *plus a third more* to elevate everyone to Western economic status.[24]

Since no one has any extra planets handy, we all have to embrace a different paradigm pretty quickly—and that's what institutions like Slow Food, EU agricultural policies, and the IAASTD are trying to do. Those yield-per-acre zombie arguments are still used to "prove" the efficacy of Green Revolution or industrialized agriculture. But knowing what we do today about the effects of industrial agriculture on soils, water, ocean dead zones, biodiversity, and climate change, it's more realistic to argue along with Bob Watson, the director of IAASTD, that in terms of agriculture, "business as usual . . . will leave us facing a world nobody would want to inhabit."[25]

Bruce Bridgeman, a professor of psychology and psychobiology at the University of California, Santa Cruz, sums up the true underlying problem, which decades of public health, demographics, and food distribution research have tried to address: "All efforts to increase agricultural production will be canceled quickly unless populations can be stabilized. The recommended reforms [in agricultural production] will help only once, but population growth is continuous. What happens when the reforms are accomplished, and population growth has literally eaten up the benefits?" Bridgeman, along with many others, is making the point that if humans don't get a handle on their numbers, anything else they may do to try to maintain the health of this planet, whether it's cutting emissions from energy use, growing food without chemicals, using less water, or protecting forests, will be futile in the long run.

The question is central, and population researchers haven't been idle. Bridgeman points out that we know exactly what to do to contain population growth, and oddly enough, it's the same thing we need to do to help feed the world. The list is often referred to as HEEP because of the first letter of each rule:

1   Health care for all, including reproductive health.
2   Education, especially for women, which is the strongest correlate of reduced birth rates.
3   Employment, especially for women.
4   Pensions that people can trust.

As Bridgeman concludes, "These four reforms are things that people want and are absolutely essential for agricultural or any other reform." When people have access to these forms of security, their reproduction rates drop like a stone, and they are more willing to share what they have.[26]

Two vastly different political states have already been forced by circumstances to grow and distribute food and to share resources in exactly this way. In the first edition of this book, we talked about how Cuba was driven by the U.S. oil blockade to become the modern world's first country fed totally on organic food. Because of the blockade, Cuba lost access to modern fertilizers, pesticides, and parts to repair their agricultural machinery; as well, 74 percent of the population migrated into the cities, leaving the countryside short of farmers. In only seven years, however, between 1990 and 1997, Cuba learned how to feed itself by paying higher prices to farmers, using modern organic technologies, establishing smaller production units, and encouraging urban agriculture.

The latter was largely a bottom-up achievement. With a little government encouragement, tiny plots of vegetables, spices, and medicinals sprang up inside every city, on lawns, in vacant lots, and in parks. Farmers' markets were deregulated so that the growers could sell directly to consumers, something that has become increasingly difficult to do in Canada and the U.S. The Cuban government also provided free assistance and advice, as well as biological control agents and access to low-cost manure and compost.

Today Cuba consumes less than a tenth of the chemicals it was buying in 1989, and the country's forced experiment has proven, beyond any doubt, that organic food production can match or surpass that of industrial production when combined with market and tenure reforms. It wasn't easy, and to this day no one is exactly getting fat. But the country's health indicators, such as mortality rates, both infant and adult, compare favorably to indicators in the U.S. and to the rest of the industrialized world; they are far better than those common in other tropical countries with similar resource bases. Organic agriculture works for Cubans because it delivers comparable yields per acre but also because food and access to it is *shared*—distributed and monitored by the state—so that enough people have access, including those who grow it. State farms were turned into private cooperatives but kept small, and the kind of government subsidies—equipment discounts, low-interest loans, and tax and insurance breaks—that Canada and U.S. give to huge corporations like Smithfield and Cargill, in Cuba go to these small operations. Their small quotas keep them from merging and bloating the way ours have. There are a few cash crops grown for export, but the focus is on local food security. Ohio State entomologist Joe Kovach, who was part of a recent research delegation that traveled to Cuba to study their agricultural system, says, "in twenty-five years of working with farmers, these are the happiest, most optimistic, and best-paid farmers I have ever met."[27]

Nobody is suggesting that such an achievement makes living in a dictatorship with a serious lack of human rights acceptable, but that's another subject altogether. What we're talking about is a critical food-producing experiment carried out on an overpopulated, underresourced tropical island, a place that has had a hard time feeding itself under any type of government. Without modern technology and fossil fuels, between 1994 and 1998, production of tubers and plantains tripled, vegetable yields doubled and then doubled again. Today most urban areas are able to provide the huge majority of their populations' food needs within the city limits, with the slack taken up by countryside crops. There are almost no imports.[28]

One side of our food equation requires producing what we need without destroying the basis of life—air, soil, and water—and Cuba has proven that can be done. The other side of our food equation requires not eating up the planet by producing too many people for the finite land and water base. Southern India, another part of the world where the population is poor and marginalized, has a big lesson to teach as well. Kerala is in the unenviable position of being India's poorest and also most populous state. However, for many years, the three major parties of this democratic state's government have all supported a very effective health program for women and children under five, the key correlate to lowered population rates. They have also ensured that women have access to education.

The state provides health care, including vaccination and free treatment of scourges like leprosy and AIDS, as well as basic staple foods like rice, along with cooking and lamp oils. Women still aren't fully employed in Kerala, but their status is better than anywhere else in India. The general literacy rate, 91 percent, is one of the highest in the world, and women are free to teach or to become doctors or politicians. Keralans can look forward to a life span less than a Canadian's but comparable to the U.S. average, and significantly, like women in the First World, Keralan women outnumber men and also live longer; a very unusual circumstance in India and in poor countries in general. The number of children each woman in this state produces—1.73—is considerably below the U.S. average of 2.1 and comparable to Canada's low 1.66 rate. Which means that even managing to institute *part* of HEEP's four rules helps people control their numbers, so they're better able to share what they have with each other.[29]

What has happened in Kerala demonstrates that when women and children especially are given some security and health and educational support, swelling populations go down. And what has happened in Cuba is a real-time experiment proving that chemical-dependent monocultures and genetically engineered seeds are not the answer to hunger. That argument was skewed from the beginning, since statistics for Green Revolution practices' impressive yields per acre never take into account the losses to soil, water, and wildlife biodiversity that these processes cost a region. They don't even subtract the dollar costs of the chemical inputs! In any event, studies running five instead of one or two seasons have revealed that industrial yields are generally high for only the first year; after that they drop below those of traditional, organic crops, because the soils are so rapidly exhausted by the chemical inputs.

Nearly all of the positive Green Revolution studies, even those coming out of universities, were paid for by chemical companies and other Green Revolution proponents, making their results highly suspect. Today, as the IAASTD report points out, even if these practices actually delivered what they claim, we'd still have to abandon them. Industrial agriculture technologies *all* require heavy use of fossil fuels. Fossil fuels cannot be the basis for future food because they destroy the atmosphere, destabilize climate, poison soils, deplete water, and contaminate the oceans. Even if we wanted to ignore that, we're almost out of cheap supplies of oil. It simply isn't in the cards that such a system, in any of its guises, and that includes biotech, can feed the poor.

BOTTOM-UP

"It is one of Africa's greatest ecological success stories... a model for the rest of the world."... local farmers have used picks and shovels to regenerate more than 19,000 square miles of land.

Chris Reij, geographer at VU (Free University) Amsterdam[30]

A recent story of subsistence farming illustrates wonderfully what we need to understand about feeding human populations sustainably. This story of simple, traditional technologies used by desperate and very hungry people to restore their lands proves that success or failure depends on two things: whether the techniques fit the particular ecosystem, however small it is; and whether the local people have control over the land's management.

Yacouba Sawadogo, a farmer living in Burkina Faso with three wives and thirty-one children, had learned about a technique called *zaï* from his parents. The zaï is well adapted to conditions in this cruelly poor north African nation, in the heart of the Sahel, where productive savanna has been converting to desert since the 1970s. For years Sawadogo spent every dry season hacking thousands of foot-deep pits into his desert fields; he placed manure in each one to attract termites, which digested the organic material and allowed it to be absorbed by the soil. He also dug channels in the cement-like earth, which loosened it up. When the yearly rains came, water would trickle through the underground labyrinth of termite tunnels, and Sawadogo would plant trees—trees that could survive in these desert conditions. Soon he was noticed by his neighbors as "the only farmer from here to Mali who had any millet" and ended up forming a zaï association to teach hundreds of others.

Sawadogo also became known for using an incredibly elegant and cheap water-retention scheme called *cordons pierreux* (rock strings), in which fist-sized rocks are laid out like chains perpendicular to the flow of rainwater. "Snagged by the cordon, rains washing over the crusty Sahelian soil pause long enough to percolate. Suspended silt falls to the bottom, along with seeds that sprout in this slightly richer environment. The line of stones becomes a line of plants that slows the water further. More seeds sprout at the upstream edge. Grasses are replaced by shrubs and trees, which enrich the soil with falling leaves. In a few years a simple line of rocks can restore an entire field."[31]

These techniques don't leave much scope for the high-tech, outside-input programs of the Gates Foundation or the products of Monsanto, but, since they were initiated and spread rapidly by the poorest of poor farmers, are right now transforming enormous chunks of Burkina Faso, Mali, and Niger into what has a chance of being permanently productive land. As always, there's a catch. Sawadogo, like all the other farmers in Burkina Faso, doesn't really own the land he works. Farms are leased from wealthy landowners who can repossess them any time they please. Unfortunately, his farm lay close to the city of Ouahigouya, and in 2008 Sawadogo saw the beautiful forest he single-handedly created annexed for timber and city lots. He gets a tenth of an acre—the size of a tiny city lot— for his decades of backbreaking restoration labor; even his bedroom and his father's grave are being sold in other lots.

One can imagine that most neighbors, seeing this, might give up on restoring their own land. In nearby Niger, however, small-holders have managed to wrest control over their farms from the government and can benefit from the work they put into them. That means the environment can benefit from the land's restored ability to hold on to water and carbon. This is the principal reason why the struggles of indigenous and traditional people for tenure over their lands, also seen in countries like Bolivia, Ecuador, and Peru, are so important to every one of us—not just to our most elementary ideas about human justice and sharing but to the future health of the entire planet.

Again and again, it is indigenous and traditional people, the dispossessed, the poorest of the poor, who not only end up defending natural systems directly but work out what are most often the best ways of sustaining and restoring them. Even when it comes to livestock, it's the traditional and aboriginal people of the world who are trying to hold the line against monoculturing our future. A 2007 Food and Agriculture Organization (FAO) conference on the state of the world's animal genetic resources admitted that diversity is plummeting at a terrifying rate—two thousand more species at risk every four or five years. Livestock diversity, like plant diversity, protects the food supply against future diseases or changes in conditions. A rare breed like the British Tamworth hog descended from a breed originally developed in the West Indies and so does very well under hot conditions, while the black, Hebridean sheep may have as many as six horns and can subsist entirely on brambles and woody shrubs, making it "of considerable value in ecological projects to control invasive scrub." But they and so many more are endangered.[32]

In the first edition of this book, we told the story of farmers in northern India, associated with Vandana Shiva's Navdanya group, who were turning down grants from the World Bank encouraging them to switch from their traditional cattle breeds to Holsteins and Jerseys. Their genetic banks, individual village bulls, have been painstakingly bred over centuries to provide what each locality needs. That's not just high milk production but the strength to pull loads and create a lot of manure, desperately needed for fertilizer and fuel. Most importantly, these cattle can survive very poor feed, a cruelly hot climate, and many diseases. The pressure to succumb can be intense; one village near Dehradun in northern India was denied a World Bank grant to bring in needed drinking water because the villagers refused to give up their native bulls in exchange. This is an example of farmers with some ownership rights who nonetheless don't have full tenure—the ability to manage their farmland as they need to. If they try to resist outside interference, they are pressured, economically punished, or legislatively prevented.

No one is suggesting that some central legislation isn't necessary to prevent farmers from dumping pesticides in streams, being cruel to their stock, or otherwise threatening local sustainability and health. But centralized regulations commonly favor cultural or economic agendas that can have damaging effects on local ecosystems. That's why local input is key, and legislating farm management is a delicate balancing act all over the world. It shouldn't be subject to centralized control, like the West's Green Revolution coercion, any more than it can completely be dispensed with, as in failed states like Haiti or Somalia.

In the West's management system, the reason why breeds like Holstein and Hereford cattle and Leghorn chickens have become so ubiquitous is not because they taste better or even because their production levels are the highest. It's because their constitutions are tough enough to survive the ghastly conditions of confinement, overcrowding, and unnatural feed that are central to industrial agriculture. Although the response of the UN's FAO has been to establish gene banks, this loss of diverse breeds is a direct attack on the very forms of smallholder agriculture that are vital to a sustainable food future. Activists are agitating for a change in policies that force industrial agriculture on farmers through government programs and slanted economic support. One such group is the Livestock Diversity Forum to Defend Food Sovereignty and Livestock Keepers' Rights.

They represent pastoralists, indigenous peoples, smallholders, and NGOs from twenty-six countries. In September 2007 they demanded a new Global Plan of Action from the UN that "challenges industrial livestock production," because "the dominant model of production is based on a dangerously narrow genetic base... propped up by the widespread use of veterinary drugs." They pointed out that "it is clear that the rights of livestock keepers are not compatible with intellectual property rights systems [i.e., gene banks] because these systems enable exclusive and private monopoly control." What they want is "livestock keeping that is on a human scale, based on the health and wellbeing of humankind, not industrial profit." With the new IAASTD report, they're a lot closer to finding powerful allies to help them toward this goal.

EVERYTHING I WANT TO DO IS ILLEGAL

Everywhere, [our] would-be clean food producers are hampered, stymied, and petrified of getting crossways of some labeling or food police. It is absolutely the most significant reason, especially in the livestock sector, why our movement has not displaced more industrial farming systems.

Joel Salatin[33]

Today, especially across North America, we're dealing with dangerous emerging diseases like avian flu, H1N1, and MRSA, as well as a resurgence of old ones, like *Listeria* and *E. coli* contamination, throughout modern food systems. This is almost entirely because of industrial practices and is happening even though health and hygiene rules abound. Any American or Canadian farmer knows that every aspect of agriculture is regulated, from manure disposal on the farm to meat-packing and cheese-slicing conditions. Mad cow disease is caused by grinding up cows and other dead animals and feeding them to cattle, to save small amounts of money in an integrated production-feed-slaughterhouse-packaging system that, despite its many regulations, did not and still does not prevent the disease's continued occurrence. The reason mad cow is still affecting Western beef, and the EU and Japan don't want to buy our animals, has to do an insistence on minimal testing and maximum "efficiencies of scale" and industry integration across Canada and the U.S. The recent *Listeria*, *E. coli*, and *Salmonella* outbreaks, which have killed hundreds of people in North America over the past few years and are a long way from being under control, also come out of our vertically integrated food system. It throws the production of thousands of farms into a centralized processing or packaging facility, so one microbe in one animal, plant, or piece of cheese can contaminate millions.

Scientists know perfectly well that H1N1 and avian flu viruses originate in the crowded, airless conditions of mass confinement facilities for hogs and poultry, which have long been seen as the perfect way to breed new viruses that can jump to humans. But governments in most industrialized countries respond by continuing to subsidize these operations with tax dollars, while making it illegal for backyard poultry to be let outside their pens. And while most state and provincial livestock regulations support and abet such operations, they also make it illegal or economically impossible for a smallholder to raise a few chickens, turkeys, pigs, or beef cattle, to keep for him- or herself, or to sell eggs or slaughter a few animals for a local market. In fact, in most of Canada and the U.S., it has actually become illegal for a few grandmothers to get together to bake pies for a school or church bake sale.[34] To be certified as clean, food has to be packaged and proven to come out of an official, government-sanctioned, and therefore corporate, probably vertically integrated, and far more dangerous kitchen than Grandma's. Yet the old-fashioned practices of backyard stock raising and casual local market selling are completely legal in the EU, where the population enjoys the benefits of the safest food in the world.

Joel Salatin would take advantage of this European agricultural model if only he could. He's a remarkable farmer who found a way to apply Allan Savory's Holistic Management principles, developed for dryland ranches, to mixed farming in crowded Virginia. He's also a celebrity, author of a dozen books on how to farm organically and sustainably in order to benefit the local community and the world's ecology. His "pastured chickens" were the first to popularize hoop houses and "feather nets" as a way of moving chickens across a landscape so that they could eat bugs and help the soil while being protected from predators.

Agro-specialist E. Ann Clark says, "I think Salatin is a genius. His great gift is 'enterprise stacking' or 'ecological energies'; that's the way he uses the natural instincts and productions of two or more species in the same barn, on the same land base. His outcomes—that is, the production of eggs, meat, even cultivation and weeding—are synergistic and impressive."[35]

Salatin raises farm stock on a 550-acre spread in Swoope, Virginia. He produces hundreds of pounds of pork and beef every year. The farm also raises broiler chickens and produces eggs from three thousand layers. Salatin's farm embodies almost every quality we've been discussing in this book: Holistic Management (he makes his goals consistent with his highest aspirations); biomimicry (he experiments with production systems that mimic nature's); and values (he pays attention to the real bottom line of happy animals, satisfied customers, and a healthy community and land base). And he realizes a healthy profit. The farm grosses $250,000 a year, which supports four adults at salaries commensurate with what each person could make in a $35,000-a-year city job.

The farm has to be viable economically or Salatin couldn't keep it going. "I'm an unabashed capitalist," he says, "But capitalism without ethics is just greed." He sees the animals on his farm not as economic units to be exploited but as partners helping keep the land and his family healthy.[36] Salatin knows about Allan Savory's methods of mimicking the patterns of grazing animals being chased around by predators in order to increase pasture fertility. He says, "On our farm we've got these areas that are all brambles, briars, shrubs, and bushes. We don't use bulldozers; we use pigs to convert that forested scrub to healthy pasture. We put thirty pigs on a quarter acre at a time; what they do at first looks awful. You'd think we'd wrecked the land. But we're providing them with the kind of habitat that they would have in nature. A couple of weeks later, the way they've torn things up has stimulated the succession of more valuable species than were there in the first place. The land evolved with this kind of disruption. So in just one season our pigs have converted the overgrown scrub to the kinds of perennial grasses and clovers we can put the cows on."

Salatin says he's read the management classics and he understands that the great plains and pastures of the world that support agriculture became endlessly fertile "not by plowing and fertilizer but because of *co-evolution* with large populations of herbivores. The antelope and wildebeest on the African plains, the huge herds on the Steppes, the buffalo here—there's always been a symbiotic relationship between herbivores and foragers." So Salatin follows Savory's basic grazing principles, bunching up the grazers so that they disturb the soil, then moving them off so that the land can rest. Calculating the amount of land needed per cow demands real skill. "It's an art, not a science," he says. "It depends on whether the cows are dry or not, the time of year, the type of pasture [but]... if you know your land and animals, you can figure it out. You'll never get the right balance if you're starting from the industrial paradigm—treating living, individual creatures as if they were dead things, numbers in your profit margin."

Salatin's brand-name broilers are called Pastured Poultry, and they get to move from place to place just like the cows and pigs. They follow the cows in order to eat fly larvae and other insect life hatching in the cow pies. This not only creates dark, glossy yolks in their eggs, it spreads the manure and takes care of any pests attacking the pasture—grasshoppers, caterpillars, and the like. He says a hundred birds will eat up seven pounds of insects a day, and he uses them to "improve the pasture, while making good chicken meat."

But Salatin's latest book is not like his first ones, about how to build hoop houses or help your family learn to love farming. It's called *Everything I Want to Do Is Illegal*. His competition in the food world is industrial chicken farmers, whose birds never see the light of day. You don't want to know what's in the mess your broiler was fed, but feed often includes the manure and freshly ground-up bodies of fellow chickens, as well as dead hogs and cattle.

Remember the first avian flu scare in 2002, when sharpshooters were waiting for flocks of wild geese and ducks to cross into the U.S. from Russia, ready to kill them to prevent them from infecting domestic birds? Ever notice how that story, and indeed all our fears of disease being spread by wild birds, suddenly vanished? It's because by 2003, serious researchers, including Michael Greger, who wrote *Bird Flu: A Virus of Our Own Hatching* and Devlin Kuyek, of the agricultural NGO GRAIN, had exposed the truth: that these viruses are emanating from the huge factory farms of chickens and turkeys that have sprung up around the world. The world's most prestigious medical journal, *The Lancet*, agreed with and cited Kuyek's impeccable research.[37] But governments still responded with legislation that increases the economic viability of industrial flocks and punishes the only possible solution: small, local ones.

Medical researchers have known for years that industrial farming conditions must inevitably lead to influenza epidemics. On hog farms the animals are in crowded, unnatural conditions, and their airways are raw from the ammonia and pathogens in the barns' air from manure stored below the slatted floor or just outside in open lagoons. Their feed is laced with antibiotics to ward off infections and to artificially stimulate their appetites, which otherwise would barely exist. Their immune systems are taxed to the maximum by living without air, sunshine, or exercise, much less the social life an animal as intelligent as a dog requires, so their keepers are acutely aware that the slightest germ or upset could lead to a chain of influenzas and viral mutations that could decimate the herd. The obsessive rules of an industrial farm are not a sign of care but of half-sick animals continuously on the brink of catastrophic disease.[38]

Whatever happens with H1N1 and its related swine flu variants in the future—all of which could jump to humans—people need to remember that it started in the industrial farm setting and developed the ability within that environment to infect humans. As long as we persist in producing meat this way, avian and swine flu viruses will continue to threaten the world population on a much higher level than they would normally.[39] As GRAIN, the NGO that exposed the unfair attacks on wild birds during the avian flu scare, says, "Factory farms are time-bombs for global disease epidemics. Yet, there are still no programmes in place to deal with them, not even programmes of independent disease surveillance."[40] If we were to include the price of pandemic management—the many tax dollars poured into vaccines and antivirals, all the hours lost at work and all the lives lost—the true cost of industrial pork would be a lot more than Joel Salatin's happily rooting, pastured variety. In fact there'd be no comparison.

Of course, organic producers like Joel Salatin are the first to admit that small doesn't necessarily mean safe. He says some of the dirtiest and sickest chickens and cows he's ever seen were raised by Third World—or sloppy U.S.—smallholders. And that's the point. Size is no guarantee of safety; proper testing and truly healthy stock are. All health legislation should take size into account, the way it has in Europe, and make sure regulations are not throwing millions of people off the land to satisfy rich industrialists. Even in the U.S., where farm labor has come to be disdained as too hard and nonremunerative, Salatin claims that one person can gather the eggs from an operation his size with only about seven hours' worth of labor a week. They will net up to $20,000 a year profit, and that income is even better for pork. Best of all, the animals get to live outdoors, have social lives, and eat what they evolved to eat.

So why do we continue with the ghastly business of burning beaks off and nailing chickens to their perches, feeding hogs ground-up cows in shiny white barns with cages so small they can't even turn over? Salatin says a good part of the answer is money and power, but it's also historical and cultural. Fifty years ago, the new industrial model set what was perceived as the laudable goal of using as little human labor as possible. "It's supposed to be nasty to work on a farm, so they try to use very few people," Salatin says. The fewer the humans, the worse that care will be; that's one of the ideas that the concept of "multifunctionality" addresses. Not enough farmers explains the need for all the antibiotics and other drugs, as well as the diseases coming out of livestock confinement facilities. Worst of all, Salatin says, "We think it's much better to be in some city apartment working a salary job." But today, "the industrial paradigm thinks that if you have lots of work, you're providing for your family—that's a liability."

The real liability is a system that works against smallholders. Many North American jurisdictions have decided to protect watercourses from feedlot contamination and erosion by demanding that cattle be fenced off from natural streams and ponds, even at on-farm waterholes. That means the only way a farmer can provide water for his cattle is to drill bore wells, for between $5,000 and $10,000 apiece, that suck up millions of gallons and rapidly deplete precious groundwater aquifers. This kind of legislation, which consumers are told will protect them, often intentionally works against the organic and smallholder methods that are the only viable form of agricultural production we will be able to turn to in the future. The laws in North America and many other jurisdictions force all farmers to "get big or die." In other words, based on all the highest-level studies reviewed by the IAASTD, these laws take us in exactly the opposite direction from where we need to go.

The best news in food is that the dangers and limitations of industrial agriculture are being resisted with incredible energy worldwide; they're being discussed in every imaginable forum, from town council meetings and dinner tables to the highest levels of government and finance. For the first time since the Green Revolution of the 1950s the industrial model is being fundamentally challenged, not only by increasingly well-organized smallholders but by the largest and most prestigious international institutions.

In 2008 Slow Food guru Michael Pollan published a long letter in *The New York Times* reminding the newly elected president, Barack Obama, that the main thing that made food cheap and forced people off the land after World War II was government subsidies to staple crops. "The chief result... was a flood of cheap grain that could be sold for substantially less than it cost farmers to grow because a government check helped make up the difference." Artificial grain prices meant factory farmers could lock up cattle and feed them with more ease than traditional farmers, who had to move cattle off and on pasture. The main energy advantage for humans in eating meat is that animals like cattle and sheep can eat grasses we can't, grasses that don't need to be plowed and reseeded with expensive, patented hybrids every year. Cattle can put that cellulose into a high-protein form we can absorb. But under industrial systems, cattle are eating food grains and using up cropland the same way ethanol does. Moreover, a diet they weren't built to handle increases their production of methane, so this system has vastly increased livestock's contribution to global warming.

If meat is artificially cheap, animals' lives are as well. A traditionally pastured dairy cow, for example, can be milked for eight or ten years and can live for fifteen or twenty. Today the dairy cows chained in megabarns and fed grain mixed with "protein derivatives," like chicken manure and rags, will produce for about two years; then they're thrown away like old shoes, ground up to feed chickens. This is cruel, wasteful, unhealthy, and stupid for everyone concerned. It could be stopped tomorrow if the subsidies on grain were removed. Pollan essentially argues that without artificially cheap grain, factory farming and all its attendant horrors—including emerging diseases—would disappear along with antibiotic use and related pollution woes.

What is coming out of the international dialogue on food is not "How can we afford to grow organically?" but "How can we possibly afford to keep growing food industrially?" Without the benefit of help of any kind (and facing an amazing number of obstacles), organic growers already produce between 80 and 100 percent the yield rate of industrial operations and are expanding at the rate of 20 percent a year. What would happen if organic growers had some favors thrown their way, or at least were on that level playing field that proponents of globalization are always talking about? Of course, there's the fact that organic and polycultural or multifunctional farming models aren't as simple to manage as the current model of "driving and spraying" corporate-supplied seeds coated in corporate-supplied chemicals. "Sun" farming, as Pollan puts it— that is, not using fossil fuels but only the sun's energy—requires more knowledge and training and more rural workers. A lot of people are asking, What's wrong with that?[41]

Many people are looking for careers away from corporate or cubicle drudgery, and a lot of jobs would open up with a new model of agriculture. Most farmers' spouses are already balancing books, doing taxes and paperwork. With fair marketing systems and better farm-gate prices, a typical farm family could hire out some of this labor and end up with a pretty good life. In North America our farmers are disappearing; the average age is 55 to 60 years old, and new farmers are not coming in. It's not too attractive, as Joel Salatin puts it, to be "a serf for an agrochemical company." But true tenure and control over crops could be regained if government regulations favored the following: production based on sun rather than fossil energy; leaving many areas fallow or natural; water conservation through management practice and crop choice; cultivation of local varieties and markets; and legislation supporting these practices. Local government agencies could take care of regulating the new local markets, packaging areas, slaughterhouses, on-farm restaurants, and other food amenities and facilities that would spring up, all of them in need of full-time workers.

Such a revolution in food distribution is already happening in Europe and even in Canada. In 2006 Quebec held a commission on the future of agriculture and agrifood across the province. A lot of people came out, from grandmothers complaining about the closure of rural schools and businesspeople bemoaning the lack of services like high-speed Internet, to agricultural producers describing their extreme difficulties in accessing local markets and their distrust of chemicals and GMOs. The commissioners were particularly impressed by young farmers protesting the industrial monopoly enforced by Quebec's only farmers' union, the UPA, which they insisted has been captured by industrial interests and often no longer represents the interests of local Quebeckers. The three commissioners heard briefs for several months, deliberated, and then tabled twenty-seven recommendations. They called for more local production and access to local markets, more help for smaller operations, and a departure from industrial monocultures. No one expected these revolutionary changes to be implemented, but in spring 2009 word got out that the UPA would have its wings clipped; it's already been downsized from two important boards. More importantly, there's supposed to be a gradual withdrawal, like Germany's, of subsidies from grain monocultures, so farmers can adjust and retool. And there may be new subsidies: $10,000 if farmers save 8 percent of their land for habitat or keep 25 percent fallow; $10,000 for each separate "multifunctionality": that is, the more varieties you produce, the better off you'll be.[42]

In short, a new day just might be dawning. Cheap food is killing us. In fact it's killing the entire planet. Real food won't increase costs much more than a percentage point or two in a family's budget, once we stop paying for the cheap "food" of the industrialists. As Michael Pollan puts it, "eating less oil and more sunlight will redound to the benefit of both."[43]

PERMANENT AGRI-CULTURE
MEANS PERMANENT HUMAN-CULTURE

Worldwide, if the organic matter in all the land that we farm and graze currently were increased by [a mere] 1.6 percent, atmospheric CO_2 levels would be at pre-industrial levels.

Abe Collins, Vermont grass farmer and permaculturist[44]

Highly sophisticated research about mineral and water cycles, combined with what we've recently figured out about the complexity of plant and animal genes, is bringing some seemingly old-fashioned but actually revolutionary advice to the agricultural debate. New forms of sustainable agriculture that are not dependent on large corporations or fossil fuels are already dealing with the life cycles of every pest, from insects to viruses and fungi, and with every possible problem, from low fertility and salinization to drought and hardpan. In higher scientific and academic circles it's considered primitive to talk about attacking an insect pest, like the cotton bollworm, by killing it with pesticides, even biologically based ones encoded in the cotton plant, as Monsanto tries to do. The scientists whose research is independent of chemical and petro-dollars have long argued that such an approach merely inspires the worm to evolve immunity to the pesticide. There's also the fact that the novel poisons and newly engineered organisms in these products can adversely affect beneficial predators and soil chemistry, which will further destabilize the environment.

A growing host of these independent academic researchers are coming to a clear understanding that soil, water, and biodiversity cycles have to be in a particular balance with one another and with the other organisms they originally evolved with, the local plants, animals, and microbes. When they are, there is no or very little need for pesticides, herbicides, fungicides, and chemical fertilizers. But most commercial and governmental agricultural agents and agronomists earned their credentials at agricultural colleges that long ago embraced industrialized agricultural methods— and pretty much nothing else. These "agricultural experts" rarely exhibit a competent grasp of chemistry, the synergy of soil and water cycles, or the complexity of genetics and ecosystems. Even though industrial farming is now scientifically proven to cause the certain decline of all plant growth and even of all life on Earth over time, the old Green Revolution methods, based on the most primitive forms of reductionist science, are still nearly all that's taught in our land-grant and agricultural colleges. That's partly because, over the past four or five decades of Green Revolution influence, any academic institution even remotely related to agriculture has come to depend on chemical companies for a huge proportion of research, institutional, and staff funding.

Joel Salatin writes about an agricultural seminar he attended, where "a PhD agronomist from Virginia Tech" told the assembled farmers how to decide what to do if they had hay and were wondering whether they should buy calves in the fall, when they're cheap, and use the hay to feed them all winter, or should just sell off the hay. The agronomist gave a complicated discussion of the "forward margin" of the market prices of storing and harvesting this forage crop and eventually advised the farmers that in a year where hay is selling at a good price, it's better to sell it and forgo keeping the calves—all that trouble storing and feeding, and then they might not fetch a decent price after the winter. "I kept waiting for him to address the value of the manure," Salatin says. Nitrogen-phosphorous-potassium (NPK) inputs—fertilizer—costs farmers thousands of dollars a ton every spring and had been the focus of another seminar, which had completely ignored the carbon (organic material) and water content in soil while harping on chemical inputs. "But this time, NPK never entered the equation," even though, as Salatin puts it, "Selling hay is equivalent to selling soil!" The accumulated hard work of insects, fungi, and other soil biota breaks down cellulose and other wastes and turns it into *virtual soil,* putting all those nutrients up into the grasses and then into the animals that eat them, just like irrigated peaches and grapes contain *virtual water* that gets exported with every container-load of fruit.

Salatin believes that "Farmers shouldn't give that stuff away to consumers for some sticker price that doesn't take the whole chemical cycle into account." He illustrates the concept by saying, "cows are the first perpetual motion machine. They eat 28 pounds of hay a day and give me 50 pounds of goodies out their back ends." Salatin considers "selling forage crops like hay... almost immoral, because it takes solar energy-produced biomass from one location and moves it somewhere else." But to the typical agricultural economist still working from a fifty-year-old paradigm, the NPK in manure derived from the forage fed to overwintering calves is *separate* from the natural cycles of carbon, air, soil, and water that created it and in his context is "without value." The only NPK source that the industrial method recognizes as being worth money is the *petrochemical* brand of NPK—the kind created from fossils, which pollutes soils and water and is only sold by big companies.

Salatin says a hundred cows naturally produce $2,500 worth of NPK-laden manure over a short Virginia winter. The old Green Revolution paradigm ignores the fact that, "Buying calves *generates fertilizer*. Selling hay *depletes the soil*."[45] Salatin is talking about the carbon cycle, one that is as important to food production as the water cycle. Every living thing is carbon based and, when it dies, is supposed to return what it has borrowed back to the Earth to nourish more life. Manure is simply dead plant life returning to the soil. When we interrupt that cycle, as we do by ignoring the organic, carbon content of soils and interrupting normal biodiversity decay, we are destroying the entire fertility process.

Abe Collins is another farmer who has learned to study the science of reality, not of agribusiness. When he and partner Ted Yandow of Swanton, Vermont, realized what their grain-fed dairy operation was doing to natural cycles, they changed to grass feeding. They saw themselves "creating a positive effect on climate change" using "permaculture" methods—perennial agricultural systems designed to mimic the relationships found in nature. As

Abe told a PBS series called *Regeneration* in 2007, "Too much carbon dioxide in the atmosphere has broken the carbon cycle," Soils don't need just NPK. "We have the water cycle and the dynamics between communities [of life]. We have the flow of sunshine through plants into the food web. If any one of these breaks, the whole system breaks... Carbon exists in the soil as fertility, and we've put it in the atmosphere, so we have too much CO_2 up there and not enough organic carbon down here, in the soil." When dead plants are not returned to where they grew, fertility is destroyed and the carbon is released into the atmosphere instead of sequestered in the ground.

Of all the people talking about the challenge of global warming and the frightening chemical changes burning carbon has wreaked in our atmosphere and oceans, permaculturists like Abe Collins are the most optimistic about what we can do to correct our mistakes. Collins says that if we stop plowing and start raising our food by permaculture methods, even partly, we can recover our atmosphere and water as well. "If you want to increase soil carbon," that is, if we want to get the carbon we've released as fossil fuels back under the ground, Collins says, "you have to have *covered soil.*" That means avoiding plowing, ripping up the root-based nutrients and exposing bare soil to air, where it will release carbon. "That's the number one condition... Decaying organic matter and living plants. Perennial grasslands are this incredible way to maintain that soil cover... [and] the fastest way to build soil carbon that we know of."[46]

A survey conducted in Central America to measure the effects of Hurricane Mitch, a devastating weather event that hit in 1998, supports what Collins says. Researchers discovered that "farmers using diversification practices such as cover crops, intercropping, and agroforestry suffered less damage than their conventional neighbours using monocultures" because their soils were in better shape and could continue to sequester carbon to nourish the next crop. The study covered 360 communities in three countries. "It was found that sustainable plots had *20 to 40 percent more topsoil*, greater soil moisture and less erosion and experienced lower economic losses than their conventional neighbors."[47]

Abe Collins points out that when Allan Savory first set up grazing procedures that help farmers keep their land "covered," the people doing it might "triple or quadruple their livestock numbers [above what Savory advised, or] maybe they'd get divorced or would buy too many things with the new money they made—things would fall apart." So, Savory came up with the principles of Holistic Management. One reason his system is not as enthusiastically championed by some of the scientists who are in agreement with the soil principles on which it rests is that Holistic Management is not just science. It's an ethical and cultural commitment. It forces managers who sign on to it to accept a wide array of principles, and they must employ "the entire equation: land, in terms of the ecosystem processes; people; and money; they all have to fit—every time they make a decision." Addressing this "entire equation" will benefit us all in the future.

THE KNOWLEDGE GROWTH SPIRAL

Lewis Grant farms 2,500 acres in northern Colorado. He has not used pesticides since 1985, and his wheat yields are consistently within the top 10 percent of yields in the state... [His six-year rotation includes] some combination of millet, sunflowers, lettuce, spinach, broccoli, beans, and hay. This constantly shifting vegetative landscape keeps disease and pest populations in check... For fertilizer, Grant uses both cow manure and cover crops... When the Russian wheat aphid devastated wheat crops throughout the area, yields on his organic farms were not affected.

Sandra Steingraber, biologist[48]

Claude William Genest, the host of the PBS series Abe Collins appeared in, was teaching permaculture at the University of Vermont in 2007. He likes to point out that the last 150 years or so have been the anomaly in terms of agriculture; before that all food-growing was organic and regional and had to take into account such factors as rainwater conservation and the carbon cycle, simply in order to survive. So it shouldn't be too hard to board that old train, this time armed with a lot of new knowledge, technologies, and tools. Permaculture is just one of many food-producing concepts that are being developed into new systems of food production around the world. It favors using natural ecosystems to grow perennials that also produce foods and products—like a modern version of traditional agroforestry.

It means that if you live in an area like southern Quebec, which evolved as forest mixed with open clearings that naturally produce and support berries, nuts, fruits, grasses, deer, rabbits, birds, and various insects, you would take advantage of that and use Abe Collins's "covered ground" concept to plant a permaculture "farm" of nut, maple, and fruit trees, interwoven with grapes, some grasses for a modest number of grazing animals, and edible perennials like borage and other greens. Permaculture operations would keep the ground covered and the carbon working to produce food, not leaking out into the atmosphere or exported as virtual soil and lost to the carbon-hydrogen-oxygen cycle. Permaculture is particularly concerned with water, how it naturally moves across a landscape and how it can be used to maximize that landscape's production. It employs simple earthworks reminiscent of the traditional bunds, zaïs, and *cordons pierreux* mentioned earlier. Where water naturally collects, a contour ditch or swale can concentrate and hold it, catching nutrients to build a productive grove of berry-yielding shrubs or hazelnuts, also directing some of the water to a little crop of squash.

Permaculture also focuses on building and road design so that the whole farm can conserve resources; but one drawback of permaculture is that proponents have been quite enthusiastic about introducing exotics—kiwis and goji berries to Quebec, eucalyptus to Africa—a practice that has led to some disasters. This part of that movement veers away from our sustainability criteria of making sure the activity fits in with the local ecosystem. The thing that's really important to recognize is that raising food this way, keeping the carbon cycle intact, is not new to humans. It took a long time for researchers to realize it, but traditional agroforestry, of which permaculture is just a recent branch, has supported large populations for centuries. There is now archeological evidence that the Haudenosaunee villages of upstate New York, for example, were not only surrounded by many square miles of corn fields but by forests that had been essentially turned into orchards. As many as one out of every four trees produced a sweet, soft-shelled native chestnut, and the mast from these trees, and the hickories, oaks, beeches, butternuts, hazelnuts, and walnuts that were also favored, attracted game of all kinds and provided flours and delicious nut creams for a human population that reached the hundreds of thousands. "Within a few centuries," Charles Mann writes in his celebrated book on pre-contact native life, *1491*, "the Indians of the eastern forest reconfigured much of their landscape from a patchwork game park to a mix of farmland and orchards. Enough forest was left to allow for hunting... The result was a new 'balance of nature,'" so successful that the Europeans didn't recognize it. And if people could attain a balance before, we can do it again.[49]

There are many more developments in farming that combine organics, permaculture, and techniques such as integrated pest management (IPM), which uses plants to repel or attract pests. For example, in order to control the invasive vine *Striga*, a parasite that poisons and strangles corn plants, scientists interplant corn with molasses grass (*Melinis minutiflora*) and silverleaf (*Desmodium uncinatum*). They crowd out the *Striga* and produce cattle fodder, while repelling corn-boring ticks and fixing nitrogen! Meanwhile, other wild grasses that host corn-borers are planted nearby to attract that pest. Turning the field back into something of a natural polyculture gives farmers a 130 percent profit on their investment, as compared with less than 10 percent when they monoculture.[50]

An especially exciting fact about permaculture, organics, IPM, and polycropping is that in the process of producing pure foods and other products, these agricultural techniques are proven to sequester carbon and foster biodiversity. These largely organic systems avoid practices that poison and kill competing species. Local vegetation is left to grow and wetlands are preserved, not just to raise fish for food but also to share with frogs, nesting ducks, and all the wigglies down in the mud. These methodologies respect agricultural and natural biodiversity. That is the ultimate test of whether we are helping to destroy or to support the biosphere.

Another defining feature of the new agriculture is that it's designed for smallholdings and isn't pushed by a vested interest like the petrochemical industry. It's a grassroots, spontaneous movement with a million heads thinking about how to achieve the same, universally desirable goal: good, nutritious, sustainable food. Yacouba Sawadogo, planting his hundreds of thousands of trees in the blazing sun of the Sahel, never heard about the permaculture principles that are so similar to his own. Prominent academics like E. Ann Clark may admire innovators like Joel Salatin without knowing he is a member of the Holistic Management movement. Allan Savory had never heard of Slow Food when he was developing Holistic Management, and although Carlo Petrini has surely met Michael Pollan, neither one knows Will Allen at Growing Power. And yet all their goals, principles, tools, and results are almost identical.

Today, farmers, consumers, and others concerned about agriculture—including the high-level academic contributors to the all-important IAASTD report and the government officials listening to rural input in forums like Quebec's Commission on the Future of Agriculture—have begun to separate cultural prejudices from real science and find what actually works. We've had to learn some hard lessons from the past to get going in the right direction, but the proof, as they say, is in the pudding. It looks like we're finally learning how to bake a sustainable one.

GREEN JOBS IN THE CITY

ECONOMIES

DOING BUSINESS WITHOUT DOING HARM

The wise and moral man
Shines like a flower on a hilltop,
Making money like the bee
Who does not hurt the flower.

The Pali Canon, 500 BC

When we wrote the first edition of this book, the idea of making a living like the bee—not damaging the world or impinging on coming generations' hopes for the future, summed up by the media term "green jobs"—was still pretty new. Today, thanks to the 2008 financial meltdown and acceptance of the reality of climate change, deforestation, fisheries collapse, and all the rest of the environmental crises, nothing could be more mainstream than the idea of "greening the economy." The verb form "greening" implies ongoing creation and is particularly important.

Greening extends beyond nurturing what we consider green jobs—manufacturing windmills, designing electric cars, installing solar panels, or monitoring forest lands. What lies ahead is to figure out ways to create green lawyers and songwriters, green teachers and hotel operators, green manufacturers, importers, parking lot attendants—and especially green financiers, making truly green money. If someone's true bliss (as opposed to their immediate family need) lies in hunting whales in the Antarctic or filling tar sand pits with petroleum ooze out in Alberta, they might not be able to green their working conditions enough to make a difference. But most of the rest of us can and are already trying. Working green and living green start in remarkably simple, unassuming ways, but they can quickly become profound.

In Chapter 1 we talked about a movement called The Natural Step (TNS), devised by a Swedish doctor who was worried about the levels of cancer in his young patients. It sets out basic principles of sustainable management that address the perturbations in natural systems caused by human activities. Some of us may be unwittingly practicing parts of it in our offices; it's behind a lot of change in the European and North American workplace. This isn't so much because TNS membership is so huge but because, like Holistic Management in agriculture and agroforestry, its principles have gradually infiltrated business thinking, even at the corporate level. Probably the most important concept that TNS has contributed to creating greener jobs is the idea of making incremental changes that are small and easy. This is called "picking the low-hanging fruit" and keeps people from feeling overwhelmed by how big the job is, if tackled all at once.

Companies can start by making sure their office snack bar doesn't sell water in plastic bottles or package food in styrofoam. They can make double-sided copies, cut down on paper, get smart lightbulbs, and arrange for some company carbon credits. And then, if they're helping to run a pulp mill with a lot of waste that's a headache to deal with and the regulators are at the door, it becomes less daunting to reach farther up the tree and consider investing in new technology that can burn that waste in order to fuel internal operations and sell energy back to the grid. The next step might be to figure out how to create products out of wastes, like making fuel from bark and sawdust. Such efforts become a lot easier when they're incremental, especially when each step brings immediate rewards, usually in terms of noticeable savings. When success follows small changes, businesses are more accepting of bigger ones, just as the people of Germany were when they reached their soft energy targets so many years early that they felt empowered to support the radical legislation of Scheer's Law.

More than ten years ago, a British teenager, Robin Foreman-Quercus, spent two years of his young life trying to protect a five-hundred-year-old oak in Devon by tree sitting—living on a small platform so that work crews couldn't cut down the tree, which was in the middle of a planned superhighway. Today Foreman-Quercus is pushing thirty. He's become a hard-working citizen in the megacity of London, far away from forests and his old life of sleeping on a few boards exposed to all kinds of weather and protesting with young friends wearing beards and dreadlocks. He's employed at a large training company that has centers throughout the country, and he's gone from being a radical activist to helping working people learn how to pick low-hanging fruit in a corporate setting.

Foreman-Quercus's company started small, installing double-sided printers across their network. Their entire paper supply is now recycled—along with their cardboard, all plastics and metals, ink cartridges, as well as their CDs, mobile phones, and other electronics, a tricky business for anyone who's tried it. Packaging is being reused whenever possible; power consumption of all kinds is monitored and reduced; the company has achieved negative carbon status by purchasing carbon offsets and promoting environmental projects. That's the low-hanging and slightly higher-up-the-tree fruit they harvested first. Stimulated by so much success, the company then invested in new vehicles that run on cleaner fuel. They instigated a policy that insists on travel by public transit where appropriate. Foreman-Quercus lived through two years of cold and rain in his ancient oak to discourage new highway construction and says, "You know, that's one of my pet issues, car travel. I've been fighting it for a long time. This year, we spent more on public transit than on car use!"[1]

Foreman-Quercus found that when greening a company, management practices are sustainable only if they produce no dangerous waste. They also have to make all parts of each business, local as well as national and international centers, part of the solution. As in nature, local areas need to have their own rules and achievements to match their smaller business ecosystem. A bottom-up approach needs to be used at least as much as a top-down approach, and setting very high goals is mandatory. So Foreman-Quercus's company formed local Green Teams based on "green champions" within the company, who help fellow employees understand and reach the key environmental goals. A Green Team in the central office keeps in touch with these local reps. The head office Green Team is composed of a staff member from each department, a further effort to localize and democratize the work.

The highest, hardest fruit to harvest is just beyond one's reach: the dirty practices of suppliers or brokers the newly green company still has to work with. If an otherwise green company has to accept a component that contains toxic chemicals, then they can't reach their own goals. That means they have to get involved in something that is fundamentally extraneous to their business. What they're engaging in are essentially nonprofit efforts: trying to spread social and environmental values to other companies. This is the real key to the rapid spread of green practices, changing what each business in the supply chain does. And as Foreman-Quercus says, at his company even this final kind of very distant and difficult fruit gathering "is slowly but surely achieving results!"

Philadelphia restaurateur Judy Wicks provides another example of business success based not on cutthroat competition but on social principle. She took her business from a muffin and coffee outlet in her backyard to one of the area's premier restaurants, the White Dog Cafe, grossing $5 million a year and employing a hundred people. Along the way she became a pioneer for local, organic food sourcing, humane stock raising, living wages for staff, community action, and international aid programs. Wicks has been featured on *Nightline* and the *MacNeil/Lehrer Newshour*, as well as in *The New York Times*, the *Wall Street Journal*, *Newsweek*, and the *Washington Post*. She was even honored as one of "Five Amazingly Gifted and Giving Food Professionals," in *O, The Oprah Magazine*, by the guru of women's interests.

After her early triumphs, Wicks expanded into something called the Social Venture Network, a way of spreading her own green business principles to other entrepreneurs. What has really taken off since then is a larger, umbrella group she helped found, the Business Alliance for Local Living Economies (BALLE). It gets small and large businesses in touch with others that also want to get greener, helping them find out how. Members learn to network with suppliers and new clients, to access grants or markets, to identify true sustainable practices, and to inspire one another. Such movements are revolutionary, even though they're nearly invisible to most of us. Members include leaders in sustainability principles like the

Rocky Mountain Institute and the Ecotrust Conservation Economy Framework; obvious institutions like land trusts, local preservations groups, and CSAs; and financial leaders like the Calvert Group (socially responsible mutual funds), Investors' Circle (aid for conscious investors), and the entire cooperative movement.

Green Enterprise Toronto (GET), which is a member of BALLE, is typical in both its invisibility to the consumer and the revolutionary opportunities it offers to entrepreneurs. As Michael Shuman, author of *The Small-Mart Revolution,* puts it, "GET is one of the premiere promoters of progressive economic development in North America. [It is] grounded in the ideas of Jane Jacobs—to create an economy that is locally owned, diversified, and community friendly." He goes on to explain that GET's sustainable business principles help "innoculat[e Toronto] from the coming risks of energy insecurity and climate change." Such organizations are working at blinding speed. Within less than two years of existence, GET was serving over three hundred Toronto members in every conceivable business—from advertising and IT support to pest control and personal coaching—with networking, web services, lunch-hour mentoring opportunities, and other means to help them assess how green they are already and how green they can become. Being a part of the BALLE network assures the sustainability of the ideas GET provides.

BALLE founder Judy Wicks says, "I created a business to be my life; to be the environment in which I wanted to live and raise my children. Since then I've realized that I literally live above the store. That's the old-fashioned way of doing business… living upstairs. Whether it's the tailor shop, corner grocery store, or family farm, you raise your family in the same place you do business."[2] That fact carries with it a community face and a community responsibility that today's international consortiums utterly lack. Having no local ties and no means of local input is one of the first signs that an organization, whether a national park, government agency, or private business, lacks sustainability and accountability.

The current perpetual-growth economic model, however, always encourages businesses to "go public" and expand away from this local ideal, to aspire to become a huge chain or multinational. This seriously flawed model developed largely because, as Wicks explains, "bigger is better" is always the push in business schools, worldwide. People accept what they're taught without considering the real effects of endless growth on a finite planet. Wicks says she moved away from this disastrous model for reasons that were at first personal, related to her fundamental moral standards. She stays private so that she doesn't have to answer to shareholders about how she decides to spend her profits.

These days, helpful groups like GET and BALLE are a mouse-click away; they'll show anyone who cares to learn that the business of making a living doesn't have to be only about making money. It could be and increasingly is becoming a way to help others and ensure the long-term viability of our natural resource bases, while helping entrepreneurs and workers earn decent livings that won't evaporate in investment bubbles or hostile takeovers.

The real key to greening more jobs and management styles is to recognize that collective human society—individuals and organizations, including business schools—invented the system that has become so greedy, perverse, and broken. That means that current business practices are not based on some natural system law. We could all get together and invent another business system—one that works to support the entire planet, not to destroy it. And in fact that's what these invisible organizations have rolled up their sleeves and started to do.

THE REAL CENTER FOR CHANGE

Each citizen has immense capacity to accept personal responsibility for the Earth.

Mission statement, Center for Earth Leadership[3]

The good news stories in this book support the idea that vital changes in social values are very often born out of the experiences and deep commitment of specific individuals. Systemic change seems less likely to come directly out of large organizations and governments. We see this in Rajendra Singh's reanimation of dead rivers, Maude Barlow's fight for our freshwater systems, Robin Foreman-Quercus's greening of his company, and Paul Watson's defiant efforts to protect ocean life. Dick Roy, who quit his job as a corporate lawyer more than fifteen years ago to found a variety of NGOs, including the Oregon Natural Step Network and the Northwest Earth Institute, says the reason is that organizations and governments generally "do not have the *freedom* to be the assertive drivers of fundamental change."

That's particularly obvious in terms of the disappointing realities of political expediency or the profit drive in business. Roy's new Center for Earth Leadership has its own take on why even hardworking educational research organizations, NGOs, and government agencies usually get only so far in convincing people to make fundamental and uncomfortable alterations in their lives, claiming "*Motivation*, not education, is the principal impediment to assuming a leadership role." So the center's programs concentrate less on educational information and more on "deepening [each person's moral] commitment to protect the Earth."[4] In short, it's a specific effort to help thousands more of us find our own endangered whale, ancient oak tree, or dry riverbed—to make a passionate commitment to see something through to the end, as we do when our babies are born. This quiet little NGO is working to create personal epiphanies that will turn more people into the kind of protective powerhouses that have populated this book: thousands of Paul Watsons, Maude Barlows, and Rajendra Singhs.

Understandably, given Roy's background, the Center for Earth Leadership devotes particular energy to greening corporate lawyers. To turn a legislative system currently abetting the plundering of the world's resources into a source of salvation for the planet amounts to bravely diving straight into the gaping maw of finance and power, although Roy does so incognito. He starts gently, helping law firms institute greener office practices—recycling toner cartridges, saving paper. Pretty quickly, however, the program moves on to training lawyers to assess *all* of the legal actions they're involved in for their long-term sustainability. Such an exercise cannot help but lead to a startling realization of what's wrong with our current system of legislation, which nearly always favors destroying the planet's real wealth in order to feed a human fantasy about endless money. The Center for Earth Leadership is already working with all three of Oregon's law schools and with many lawyers in other states. We can expect, perhaps, the same kind of sea change in Oregon law practice that has affected local business and government, and has made Portland a national leader in sustainability.

Much of this change is thanks to the innocent lunches Dick Roy organized in the early 1990s with CEOs, local businesses owners, city planners, and government bureaucrats. In fact it's because of Roy's early work with Nike, headquartered in Portland, that middle managers there found the courage to make the sportswear giant the first major company to ban polyvinyl chlorides, against tremendous pressure from the vinyl industry, and work to remove

357

hormone-mimicking phthalates from their products. Today most national governments, even in recalcitrant Canada, are following suit. Nike didn't have to reinvent the wheel to comply with the TNS principles that Roy convinced them to adopt; this is the same process, in reverse, that made dangerous chemicals and other human-made materials so ubiquitous in our lives. Huge companies spent research funds to develop PCBs, flame retardants, and all the rest, and subsidized their early manufacture; that brought costs down and made the new chemicals desirable for others. So finding sustainable products and processes works the same way. Once you get a few of the big producers on board, the whole landscape will keep changing, just as it has with the availability of organic foods.

Nike's first director of sustainable development, Sarah Severn, was brought over from England expressly to make sure that Nike could comply with the stringent new EU rules on chemicals and retain their European markets. That purely economic motivation, helped along by Roy's lunches, inspired individuals in the company to change their corporate culture in a positive way. Sustainability often begins this way, as a marketing decision; but it quickly begins to have cultural repercussions throughout a company. Nike's NEAT (Nike Environmental Action Team) also got this huge corporation to switch to organic cotton for a mere 3 percent of its needs. But with their global market, this policy singlehandedly changed the entire climate for organic cotton growers. Today NEAT has reached out to all of Nike's suppliers and subcontractors, trying to green its entire supply and distribution chain.

Of course, there are limits to the change possible within most business organizations. Despite their wealth and power, publicly traded companies are far from free to do as they wish. As Roy recognizes, these companies have a fiduciary responsibility to their shareholders. They are legally forbidden to engage in activities that help society or the environment *per se*; they may do so *only if such activities do not in any way interfere with making money*. So with them, sustainable solutions can only be a sideline that will stop abruptly if their efforts impinge on the bottom line. Still, it was local businesses that funded the free downtown transport by light rail in Portland, because they saw it as ultimately good for sales. As Dick Roy says, at such businesses and at Nike, "They're pushing the envelope of what you can do for sustainability within the corporate structure." Outside of it, even more can be done.

Smaller, private businesses involved in any kind of local manufacturing, along with service and financial ventures operating either as private companies or as cooperatives, are in the best position to create real sustainability revolutions. If they want to go green, the first thing that businesspeople need to recognize is that they will eventually need to organize their businesses in two basic ways— quite apart from and in addition to any efforts they've made to recycle, cut energy use, or subsidize employee transit. Truly green businesses must commit to a particular locality and embrace transparency, local investment, and responsibility. And, they must stay private and avoid becoming a publicly traded corporation. Bitter experience—the Body Shop leaps to mind—has shown that even the greenest businesses lose control over many of their ethical goals once they accept the rules of public trading.

Legislators haven't overhauled the system of how corporate charters are granted since the late 1890s. The fact that lawmakers back then passed serious restrictions on the corporate concept means that proper legislation can help society to break the chains of the current one-goal (profit-only) business model. There are a great many NGOS now devoted to the work of reinventing corporate charters.[5] However, until these groups succeed and business goals become subservient to our planet's real survival needs, instead of vice versa, businesspeople who are really committed to the Earth have to consider keeping their companies private. That doesn't mean they have to be tiny, unprofitable, and insecure; quite the contrary. A lot of local, private companies and business co-ops can enjoy economies of scale or access to markets by working together as an industry, under umbrellas that work essentially like GET or BALLE. Or they can manage themselves so well that the world will come to them.

The Collins Pine lumber company, discussed in Chapter 4, is a good example of a green enterprise that is also profitable. It grosses about $250 million a year but has retained its private status along with its local, family values. Its story makes the point that growing more excitingly complex within and achieving long-term financial success without can come to businesspeople *because of*, not in spite of, their sustainability focus. Keeping a business eye on values that go beyond profits also provides jobs to be proud of, where people can push their skills and support their families while making a moral contribution to society.

Barry Ford, recently retired as Collins Pine's head forester at their Almanor Forest, was in charge of making sure the forest was healthy, which, to any well-trained forester, should mean that it's absolutely bursting with diverse plant and animal species. Ford rejoiced in the fact that "We have the great gray owl feeding in our meadows. There are only a hundred pairs left in California; we're the southern end of their range, and because of that, the regulations stipulate that we have to leave a 600-foot-wide strip of trees around all the meadows." He also was proud that this economic disincentive for the company did not require him to abandon his values as a forest professional. "In order to have those trees available for our use, I could have said there were no owls, and frankly I haven't seen any. But I know that's the right habitat, so I've said yes because they could be there now, or if we save the place, they could be coming."

There's a moral component to green jobs that researchers have discovered means as much to people as profits. Vice-president Wade Mosby says, "I'll be sixty-two next fall, but I don't want to retire. The satisfaction of knowing you've improved things is really important. I saw my dad clear-cutting these forests, destroying the places he loved, that he depended on for a future. [That's why] I waited for years for a chance to work at Collins Pine. That's typical of many of the people who work here."

Such economic disincentives would cripple a publicly traded company in Collins Pine's situation. Its entire 298,000-acre holdings take a hit from government regulators for the very reason that its practices are so good and its forests are more ecologically rich. "Because we're good managers and have more species, we get more stringent regulations on our lands to protect the fish and game species we've managed to bring back. You'd think that kind of practice would earn a company tax rebates or some other kind of encouragement, but it's just the opposite. You have to shoulder economic and regulatory liabilities if you also want to protect the resource for the future!" And yet, like organic farming—despite no advantages and considerable disadvantages—Collins Pine still remains the most economically viable lumber company in its part of the country. One can only imagine what would happen if more businesses were encouraged to operate the same way.

HEAVEN AND PURGATORY

As a co-op developer, making the pilgrimage to Mondragón was like I had died and gone to co-op heaven.
Margaret Bau, visitor from Cleveland, Ohio[6]

We are not angels and this is not paradise.
Mikel Lezamiz, Otalora, Mondragón training cooperative[7]

While individual initiative is vital to the success of humans, cooperating within a group is equally important to the survival of any endeavor. In fact, when push comes to shove, say in a war or other crisis, cooperation is more important. It's ironic that right-wing political parties talk about rugged individualism and independence, while the left wing talks about cooperation and the suppression of individual desires for the benefit of the common good. Ironic, because for the last twenty years, it's the right side of the political spectrum that has managed to master internal cooperation most fully, to present itself as one voice, say of economic globalization or the necessity for private and not public management. Meanwhile, the center and left have remained, especially in North America, rent by internecine quarrels, unable to even meet for discussions, much less cooperate. The "spoiler" elections, involving Ralph Nader and the Democrats in the States and the Liberals, N D P, and Green Party in Canada, as well as Barack Obama's problems with his own party, have demonstrated that few politicians to the left of George W. Bush have been able to cooperate effectively with their fellows.

Most green business ventures are created by inspired individuals like Judy Wicks or Truman Collins; but to realize their goals, their work *has* to shift to the cooperative and collective level. Individual green activists like Maude Barlow and Paul Watson have certainly taken personal responsibility for the Earth, but they have also built huge, international cooperative organizations to turn their personal commitment into on-the-ground, collective reality. Like most environmental NGOs operating today, the Council of Canadians and the Sea Shepherd Conservation Society rely on a mixture of grassroots democratic and top-down corporate management. But there's another business model, besides the top-down or the bottom-up ones of corporations and NGOs, that's much more common than we realize: the cooperative. In North America, we mostly think of co-ops as small nonprofits that might provide organic food or run a used clothing store. But the co-op model can work for any kind of business imaginable, from construction and fisheries to laundry services and banking. Today the biggest ones have turned much of northern Spain and northern Italy, not to mention nearly all of South America, into showcases for what can be achieved with cooperative, worker-owned, and worker-managed business, and what directions truly green business finance might take.

The Mondragón Cooperative Corporation in the Basque Country of northern Spain is certainly the largest and most successful of cooperative business models. The region of its birth was part of the anarchist movement in Spain that was brutally suppressed by Francisco Franco in the Spanish Civil War and also by the Communists, who resented the fact that they couldn't control it. A young priest named José María Arizmendiarrieta arrived in 1941 and decided that economic development would be his contribution to the war-torn local society. The Basques had a long tradition of self-help and cooperative work, and after starting with the student-managed Polytechnic School to train the workers, Arizmendiarrieta set up the region's first co-ops to manufacture heaters and cooking appliances. Cooperatives, like all alternative management methods, are not easily recognized by banks and other conventional financial systems, which are still set up to serve nineteenth-century oligarchies and corporations. So a co-op's biggest problem is usually getting initial capital and subsequent financial services. Mondragón had to create its own credit union in 1959 to allow its members to access finances and provide start-up funds for new ventures. Today over 100,000 people are directly employed in 160 large cooperatives, doing mostly manufacturing work. Eroski, the largest Spanish-owned retail food chain, is a Mondragón cooperative. Statistics show that these businesses are *twice* as profitable as the average corporation in Spain, with employee productivity surpassing any other company in the country.[8]

Mondragón turns over in excess of $5 billion a year through electronics and linen companies, auto parts, and appliances; it runs a bank, research institute, entrepreneurial division, insurance and social security divisions, several schools and Mondragón University, and health maintenance and health insurance cooperatives, all "dedicated to the common good." It has enshrined some solid social values, donating 10 percent of profits to charities, with 40 percent retained to provide an account for members. The 50 percent remaining profits are open for use by owner-employees, who pay interest rates only a point or two above what the account earns. Individuals may not start a Mondragón business; several friends are what's preferred, since the "natural bonds of friendship" are considered to be a better foundation for long-term success.

Despite Mondragón's bedrock ideals, in which "people are given priority over capital," it is probably the most business-oriented cooperative venture in the world.[9] When faced with globalization and the new environment of the European Union in the late 1980s and early '90s, Mondragón chose to compete directly with conventional multinationals, joining the economic system of ever-growing profits, a decision that has resulted in many paradoxes. As it expanded its businesses beyond the borders of the Basque Country, only a third of its employees have remained worker-owners. A somewhat snide article in *The Economist* in March 2009 assessing the co-op model's ability to cope with recessions pointed out that, "People have been hired in far-flung places, from America to China; [Mondragón] now has more subsidiary companies than co-operatives... when recession bites, non-member employees suffer most... Like capitalist bosses, the Mondragón co-operativists must, indeed, occasionally handle strikes and trade-union trouble."[10] The Mondragón website claims such imbalances will be seriously addressed by 2010, when it hopes to bring co-op membership back up to 75 percent.[11]

As Mondragón's story illustrates, cooperatives can descend from heaven to hell—or at least to purgatory. If co-ops get too big and powerful, they tend to lose their democratic character. Like big labor unions, "Large co-ops become unaccountable to a disorganized, complacent membership, moving away from worker control toward conventional capitalist practices." Mondragón's difficulties show that even with an improved business model, *there is no sustainability so long as your goal is constant growth.* Trying to fit into an endless growth model that does not reflect natural system capacities always brings with it a loss of local contacts, both with people and ecosystems. Mondragón has tried to address this problem by setting up "cooperative congresses" to figure out how to deal with the massive challenge of globalization; however, its key competitors in its production of machine tools, car parts, and appliances are multinationals, and Mondragón couldn't see how to get into other businesses fast enough without competing with these companies head-on. This means that it's made the fundamental error of adapting the multinationals' unsustainable model. As its local supply chain extends to Thailand and other poor countries, Mondragón is losing its ability to demand the same values from its suppliers, values that are central to greening any corporation. Its story illustrates that as long as they don't address the goal of "endless growth," industrial co-ops can also only go so far in creating truly green jobs.[12]

What this and other examples teach us is that regardless of a business's legal structure, it has to stay small and local, at least in terms of management, in order to hold on to any values it may have, whether they're social, environmental, or even religious. It's important to note the one big advantage of the cooperative model, which is that, unlike conventional corporations, it can set up business requirements that are based on values other than profits. In the case of Mondragón, the values were social equity and job security, but they could just as well be keeping a small carbon footprint and helping maintain natural systems. By supporting and paying attention to all the new approaches people are using to green their industries, their jobs and their cities, we can learn to recognize who is really operating, and cooperating, sustainably—and heading toward truly desirable goals.

BUSINESS REVOLUTION ON THE BANKS OF THE O-HIO... AND BEYOND

Before, the managers were paid thirty or forty times what the workers made... Once they got rid of the owners, the factories began to be viable.

Diego Rosemberg, member of Lavaca,
an Argentine communication collective[13]

There are better ways to deal with competition and size than Mondragón's model, but they usually depend on government help. In northern Italy a similar cooperative movement, one that is tied to Slow Food, profiled in Chapter 7, enjoys protected status under the Italian constitution. The constitution recognizes the social contribution of cooperatives and promotes them through legislation, classifying them as nonprofits so that they pay no taxes—so long as they reinvest their surpluses in future job creation. In other words, Italian co-ops can't get huge or make old members rich. They have to keep bringing in new people, which creates more small, local centers; they're set up to help people to share.

The Italian co-op movement is nearly ten times larger than Spain's, with a quarter of a million members, as well as many more workers. They've depended less on heavy industry and more on quality production for niche markets—food and furniture, and also customized industrial parts and special delivery services. By banding together in flexible manufacturing networks, small to medium-sized manufacturers help each other establish contacts and find clients; they also "preserve the small plants where robust democratic decision making is more likely." However, even a 2,500-member co-op in Bologna can elect delegates to represent their group in bigger ones, the way proportional democracy works in governments. So up to a point, "size can be managed by how the numbers of people are organized."[14]

Many analysts feel that in the U.S. and Canada, the Italian model is one to follow, although so far there's been no broad government support. Even so, we don't realize how many cooperatives already surround us. Anyone from the U.S. knows the famous dairy brand Land O'Lakes. It's a farmer's cooperative, which incidentally did very well in the 2008 downturn because of its cooperative status. More than 30 percent of food brands in the U.S. come from farmers' cooperatives, including such household staples as Florida's Natural Brand Premium Juices, Sunkist, Ocean Spray, Organic Valley, and Welch's. Ace and True Value hardware stores are member owned and operated; *Mother Jones* and *National Geographic* are member-owned publications similar to co-ops.

In Ohio the worker-owned idea is expanding into a wide network of businesses called the Evergreen Cooperatives, which has a double mandate of not only helping unemployed or underemployed inner city residents find stable jobs but also making sure those jobs are green. Cleveland is deep in the rust belt, one of the five poorest cities in America. This failing community and its steadily falling population was built upon the worst excesses of First Industrial Revolution production. Today it's on the margins of security, seeking solutions by turning to the Second Industrial Revolution model outlined in Chapter 1. The Cleveland Foundation and the Greater University Circle Initiative are supporters of Evergreen and have started their own brand of urban renewal, using a basic Holistic Management strategy: looking not at the region's weaknesses but at its strengths. They realized those lie in the "anchor institutions": Cleveland's university, as well as its health care and cultural organizations.

The usual urban renewal efforts involve trying to attract outside industries, typically with no ties to the region, via tax breaks and lax labor and environmental standards. But the Evergreen strategy is staying local and creating homegrown, worker-owned businesses with high labor and green standards. A partner in this strategy, ShoreBank Enterprise, helped raise money for the first business, a $6 million industrial laundry in the heart of the city. The laundry is being run by fifty owner-employees drawn from low-income neighborhoods and recruited through churches and other community networks. Opened in June 2009 the building is LEED silver certified—that means built to high environmental standards, with energy-efficient driers and washing machines, as well as processes that reuse waste heat and water. After seven years on a job with high security and health benefits, employees will have built up ownership stakes of as much as $65,000 each. That should spur loyalty, hard work, and, over the long run, a far more prosperous community able to afford better housing, build better schools, and provide a better tax base for city services. Using a local policy similar to Italy's national one, as each new co-op comes online, Evergreen businesses will dedicate 10 percent of their pretax profits to a development fund that will build more cooperatives; so fifty jobs will become five hundred and then five thousand. These co-ops will grow, not in terms of the size of one business but in terms of the proliferation of their business model—which seems to be the most sustainable way to do it.

People are beginning to recognize the folly of having only one system of markets, finance, and business. Because of the 2008 downturn, however, many cooperatives are amalgamating. Many farm co-ops in particular, such as Land O'Lakes, have done so recently, even merging with conventional corporations. This is no time to let our proven alternatives be eaten up by the same old corporate model. Is there a way for alternative businesses to get big enough to protect themselves but not fall into the corporate trap? The answers might be found in South America, where a massive business revolution—invisible to most of us in the north—is well underway.

European immigrants first brought the idea of cooperative businesses to South America. It meshed well with the indigenous peoples' familiarity with collective work and resource sharing, and it received, ironically, a boost in the 1960s, when the U.S. Alliance for Progress promoted worker-owned cooperatives "in an attempt to isolate growing support for Cuban-inspired guerilla movements." Subsequent U.S.-backed dictatorships did their best to dampen the enthusiasm for this model, but when the terrible economic crash of 2001 hit Argentina, Uruguay, and Brazil, thousands of factories in the region closed their doors. Their managers and boards said they were too small, badly capitalized, and labor intensive to remain globally competitive and stay open.

Undeterred, the disenfranchised workers at these companies, especially in Argentina, broke into the boarded-up factories and turned them into co-ops and worker collectives. "We began to say, the buildings are here, the machines are here, and so are the workers. The only thing missing is the boss. Let's continue to produce; and that's what we did," says José Abelli, cofounder of Argentina's National Movement for Recuperated Businesses and president of

FACTA, the Argentine federation of self-governed worker cooperatives. "During the 1990s," says Diego Rosemberg, "in Latin America,
they said that the problem of economic development was above all
else the cost of labor... What the workers have discovered with this
type of [cooperative] organization is that the true cost that made
these businesses unviable was *the cost of the owners*."[15] This realization spread, and people began to form new alliances in every
imaginable industry. Around the same time, MST, the landless
worker's movement in Brazil (Movimento dos Trabalhadores Rurais
Sem Terra), began to organize rural worker and producer co-ops to
sell their products through agrarian reform stores. This helped support their core efforts to get land tenure reform across Brazil.

The widespread South American alliance between urban and
rural workers is also seen in northern Italy and is probably the most
vital component in the creation of truly green jobs. If urban workers do not support rural workers, the latter are left without political
allies and are generally overwhelmed by business interests that
see farm and forest landscapes as extraction and dumping sites.
Because it is the rural areas that supply all cities with clean water,
air, and food, seeing the two areas as completely separate, as we
normally do now, undermines the efforts either area makes toward
resource and eco-services sustainability.

An article on South America's co-op movement in a recent issue
of *Third World Resurgence* points out that the cooperative movement there spans not only urban and rural but all types of work.
"Each of [their] experiences is unique, shaped by the home country's diverse history." So Uruguay is best known for its housing
cooperatives, which were a factor in the overthrow of a repressive
Uruguayan dictatorship in the 1970s; "Argentina is known for its
service cooperatives and recuperated factories; Paraguay has a rich
tradition in savings and loan cooperatives; Brazil has important

numbers of worker collectives and agricultural co-ops." Venezuela, has, of all things, vibrant cooperative funeral services and is now branching out into worker collectives. Less than ten years since this movement was kick-started by disaster, it's beginning to integrate production and marketing branches—the way Land O'Lakes and the Mondragón co-ops have done—but in a completely different spirit.[16]

Cooperatives across the continent are grouping together in small, local networks, medium-sized national ones, and cross-border international umbrella groups. An example is Justa Trama, a local network of Brazilian cooperatives, recuperated businesses, and home artisans making organic clothing. They've managed to integrate every step of their supply chain, cutting out corporate intermediaries. Univens, one of their smallest members, with twenty-six workers, produced fifty thousand bags for the World Social Forum in Porto Alegre in 2005. The only way they could do that was to join with thirty-five other textile co-ops and similar businesses in their part of Brazil. Without corporate distribution overhead, the co-op's unity with others enabled it to get twice the price its bags would have fetched in the open market, and its client still paid less per bag than they had at other venues. Inspired, Justa Trama brought in organic cotton farmers, thread-spinners, and weavers in the rural-urban configuration that makes this movement fundamentally sustainable. Today, "Every step of the supply chain is in the hands of local producers, cooperatives, recuperated or small businesses." Seven hundred workers are being supported by this integrated organization, producing 10 tons of organic cotton clothing a year for markets as far away as Europe. Rather than increasing the size of the individual co-ops, they've increased their ability to access markets, all without having to deal with multinational corporations or adopting their business model.

Part of this growth in alternative business models is due to the unsustainable power of oil money. Venezuela has helped fund the co-op movement with fairs and teaching sessions, and some commodities are being exchanged across borders, like buses produced by a cooperative in Venezuela, tied to a trade with Argentine coffee producers. There's still chaos and a lack of coordination in many places, with rural cooperatives in particular competing with each other for market access. There isn't yet the kind of state support for cooperatives that Italy's constitution has bestowed on theirs. And, just as elsewhere, many cooperatives are still tempted to adopt a corporate model as they grow bigger.

For now, however, it's no small accomplishment that across the entire region, this alternative model has enabled landless peasants and laid-off factory workers to take charge of their own livelihoods and create stable, decently paid, and, most of the time, sustainable green jobs as well. These benefits to the individual workers and their families radiate outward. Reborn factories mean reborn housing and revitalized neighborhoods; and that has enabled entire cities to recover from a devastating economic recession and to expand government services.

GREENING THE CITIES

Make the community able to invest in itself by maintaining its properties, keeping itself clean (without dirtying some other place), caring for its old people, teaching its children.
Number 11 of Wendell Berry's 17 Rules for a Sustainable Community[17]

As we noted in Chapter 1, municipal governments all over the world are leading the growing movement for sustainability. Organizations like the Mayors' Hemispheric Forum and the c40 Cities Climate Leadership Group are simply bypassing presidents and prime ministers like George Bush and Stephen Harper, who dragged their feet on climate change legislation. It's surprising to see who's really getting on with the business of hard goals and multilateral agreements. But it makes sense if you know the rules of sustainability—that the best solutions will always come from the less powerful, from people on the ground, and from local communities that are closest to individual ecosystems and most aware of their needs. In the world of elected governments, that's municipalities.

Today both the city of Portland and the state of Oregon are recognized as having some of the most progressive environmental legislation on the continent—legislation that helps jobs, businesses, and cities go green. Portland has managed to make itself probably the most car-unfriendly city in car-crazy North America, without angering residents or slowing down its tourism industry. That's because it approached the situation with carrots instead of sticks. Its methods have won over even hard-core motorists with an excellent (and free) system of light rail in the downtown core. Large parking lots are located strategically outside of and around town to make the transition from car to transit more attractive, and the trains come so often that a laden shopper need hardly wait three minutes. The city's narrow, tree-lined streets give electric buses, bicycles, and pedestrians priority in two lanes out of three. All these friendly nudges have inspired both residents and visitors to leave their cars and cycle, walk, or use public transit. Not surprisingly Portland has one of the most beguiling downtowns in North America, filled with friendly pedestrians enjoying the shops, restaurants, and pretty side streets, and especially enjoying the fresh air and lack of traffic noise.

Less than a decade ago, car sharing, which we had found only in Europe and Portland, was a new way for urban residents to share a fleet of small and fuel-efficient cars for an annual fee. Today car sharing has spread all across North America. In addition many cities also have partially free bike services, with tough city bikes parked in strategic shopping, school, or entertainment areas, so you can leave downtown, ride to your girlfriend's, drop off your bike, have a long dinner, and get yourself home by grabbing another bike after the buses have stopped. Montreal's service is so popular it was expanded almost immediately.

Even neighborhoods within cities are beginning to take control of their surroundings. In the 1980s the Dudley Street neighborhood in Boston had become one of the worst urban war zones in the U.S., a result of racism, crime, and neglectful and absentee landlords engaging in speculation and using arson to collect insurance. In the 1970s realtors had forced minority white homeowners out of the neighborhood, even orchestrating break-ins to do so, in an effort to play cards with properties being sold at a fraction of their value. When collapse followed, the scammers passed the costs of foreclosure on to the federal government, and Dudley became a wasteland of abandoned houses and vacant lots—hundreds of them.

The remaining residents didn't want to be cleared out for "urban renewal," however, so they organized the Dudley Street Neighborhood Initiative (DSNI) and started gathering the low-hanging fruit, beginning with a simple campaign to prevent garbage dumping in their burned-out, vacant lots. That success quickly led to the creation of a comprehensive revitalization plan, presented to the city of Boston only two years later. Boston accepted it, making Dudley Street the first community group in the country to "win the power of eminent domain to acquire vacant land for resident-led development." Absentee-owned properties were seized and awarded to the new Dudley Neighbors community land trust. Residents got to choose to stay and develop their vacant lots as homes or gardens, but the speculators from outside lost theirs. DSNI proceeded to rebuild this vacant land with low-income housing, and created a commons, green spaces, a farmers' market, playgrounds, and more community services.

When the sub-prime mortgage crisis came along in 2008, this thriving community was able not only to withstand it but also to help homeowners beyond its borders avoid foreclosure. Dudley lot owners aren't allowed to gamble or speculate with their land trust properties; they have to agree that all future sales will be made to a low- or moderate-income buyer and follow a resale formula that permits price appreciation—but at a modest and sustainable rate. These restrictions also get them a lower property tax rate from Boston and allow DSNI to support tenants' rights in apartments. Today the initiative is working with the city of Boston to buy and repair foreclosed properties and add as many of them to the land trust as possible.

The social effects of such policies are profound. Kids who grew up in Dudley have become community leaders, politicians, and teachers. John Barros only left his home neighborhood to go to Dartmouth; now he works on the DSNI board. He says the main "sign of progress [is] that so many kids move up [in society] and stay in Dudley." And Dudley is spreading its model, showing other inner-city neighborhoods how to forestall "the storm of predatory lending, fraud, and foreclosures sweeping the country." Its approach, countrywide, has seen only two foreclosures in a sample of more than three thousand land trust homeowners.[18]

As we've emphasized in examples throughout this book, Europe still has the greatest number of truly green cities. What European cities have in common is citizens who have all agreed that their ecosystem needs, as well as their garbage and wastes, must be dealt with; and the efforts they make will enable them to have decent lives for a long time into the future. Their efforts are causing neither personal suffering nor national economic problems. Despite them, or arguably because of them, Germany and other countries with exemplary green cities are among the most prosperous nations on Earth. Striving to do what European cities do already has to become the norm across the planet because it's the right, just, and most prosperous way to build the future. Painstaking studies undertaken by the Wuppertal Institute in Bonn in the mid-1990s demonstrated that real, equitable sharing of the planet's natural resources could enable every starving, sick human on Earth to live just as well as middle-class Germans did back in the early 1970s. That is, we could all share a future with comfortable, private homes, good diets, nice clothes, and electronics. With serious efforts toward steady-state economies and populations, the only real sacrifices would be less per capita in terms of private transportation, luxuries, and waste. That doesn't sound too bad for anybody—even those of us currently taking more than our fair share.[19]

LIMITS TO GROWTH

These were very poor people; subsistence farmers with a few cattle, living on crafts like cuckoo clocks and what game and fish they could get from the forest.

Konrad Otto-Zimmermann, head of the International Center for Local Environmental Initiatives, on the inhabitants of the Black Forest[20]

It's clear that somewhere in the EU there are epicenters of individual and collective efforts that envisage a future a lot brighter than what Canada and the U.S. are setting up. These efforts most often come directly out of the past. One nexus for the kind of environmental responsibility and deep love of nature that has enabled a majority of

Germans to serenely support Scheer's Law and use low-flow toilets probably lies in Germany's fairytale region of gingerbread cottages and pine-covered cliffs: the Black Forest. Konrad Otto-Zimmermann is now director-general of the International Council for Local Environmental Initiatives (ICLEI). ICLEI is an international NGO that tries to help people develop sustainable initiatives and implement the "green" principles as set out in Agenda 21 of the first Earth Summit back in 1992, which all of the UN's member states signed on to. When we interviewed him, he was in Freiburg, the Black Forest's largest city, although he comes from Swabia. We asked if he could explain what makes the Black Forest the birthplace of movements like the world-famous electricity cooperative in nearby Schönau.

The revolution of 1848, the first movement toward democracy in Germany, found its most passionate adherents in this area, Otto-Zimmermann says, and "It's been a hotbed of democracy ever since." He explains that the concept of *Heimat* or *Heim* originated in this part of Germany. It means "home," but it also implies everything that home can represent: its culture; its natural surroundings like animals, plants, and hills; its smells; its people and buildings; its spirit. "You are part of it, and it is part of you," he says. *Heim* also has to do with democracy, "being master of oneself, no king, no duke, no one from the outside telling you how to live"—all ideas that would resonate with any aboriginal or traditional person on Earth.

Freiburg is one of the jewels of Germany, with much of its medieval town—the ancient walls, the cathedral, the surrounding squares and half-timbered houses—still intact. Five fingers of forest reach down into the town from the hills, and there is no urban sprawl—none of the spilling of tract homes and malls across the landscape that one sees in most of Germany and the rest of the modern world—because the city zoned against it. Much of the town is off-limits to cars, served by shiny electric trams that are constantly running to and fro. Even the crystalline mountain streams that used to flow through the city, filling horse troughs and feeding fountains, still sing merrily as they run along the meticulously constructed, debris-free stone canals that line the pedestrian streets and walkways.

Freiburg sits on the Rhine, one of the most navigable rivers in the world, with France and Switzerland both within shouting distance. Otto-Zimmermann worked for a year in Stuttgart, which is not very far away. He says that city is devoted to industrial development, always looking for more: "It's a survival mechanism for them—'How can we let any opportunity pass by?'—and they're always focused on business, business." Stuttgart is the big port on the Rhine, "but if you look at the map, Freiburg is better located, right on a line of transportation that stretches from Scandinavia down to Spain. They could have gotten so much money to develop. Even Daimler-Benz, which comes from here—the city turned them down!" At first, Otto-Zimmermann wondered why the region had been so resistant to development for so long, but then, he says, "I discovered that people have a certain regional mentality which leads them to protect this home area—their *Heim*. It makes them very conservative and very cautious about changing what they already have, in the very best sense of those words."

As we've seen, the democratic expression of a region's social and political values is its smallest units, its municipalities; but in Canada, even if they're long established and well run, municipalities can be wiped off the map if the larger governing body, the province, suddenly wants them to amalgamate or disappear. Moreover, each Canadian municipality, however rural or tiny, must have an "industrial zone" and is forced to accept into it any industries deemed desirable by the province. This is why the judicial wins in Quebec's Elgin and Hudson cases, described in Chapter 5, to gain the right to forbid the use or dumping of toxins, are so important to the future of democracy in all of Canada.

These rights we're fighting for are already enshrined in German law. There the state authority cannot override decisions made at the local level if those decisions are more, rather than less, strict in terms of the conservation of resources or the protection of human or ecosystem health. If a town or village doesn't want a factory or dump in their town, they don't have to take it, just as in the Hudson and Elgin rulings. However, if a municipality does want to permit industrial development or investment, that has to be approved by the regional authorities so that certain levels of waste or pollution will not be breached. "It's a system of checks and balances that doesn't prevent all bad decisions," says Otto-Zimmermann, "but certainly helps avoid them. You can't supersede a democratic decision by a nondemocratic—that is, an administrative—one. Local self-government is pretty secure here." In other words, if the state in Germany had wanted to merge surrounding municipalities into megacities, the way Quebec and Ontario have done with Ottawa, Toronto, and Montreal, they would have had to go through a public, democratic vote in Parliament. "And that," says Otto-Zimmermann, "would be very politically dangerous for the supporters of such a thing. Development, amalgamation, certain kinds of industry, they cannot be imposed."

The protection of local autonomy is only one way in which the citizens of Freiburg show their expertise in conducting democracy. They practice it in much the same way North American native groups, river-restorers in India, and holistic managers everywhere do: through the extremely long and painstaking process of "talk,

talk, talk"—achieving complete consensus. Otto-Zimmermann says, "The no-car-zone idea first came up in 1975, ten years ahead of anywhere else in Germany. As it moved through all the studies and the committees, the city required every decision to be unanimous. There were visits from all the people affected, the city took every kind of complaint seriously, they talked to the neighbors, they made sure everyone on both sides of the issue understood everything—real participatory planning." Echoing Rajendra Singh, he says, "It took *so long*! But when the decision to ban cars came, there was a real consensus. Everyone seems to love it now." He says one reason these rulings work is because the officials have the same kind of commitment to the area that the citizens do. "In city government," he says, "I noticed people from all the different offices talking to each other so carefully; they really understood the will of the developers, they understood both sides. But they would come up with every reason you can imagine *not* to develop. 'That's not the best place... maybe over here, or over there... maybe, you know, not here at all.' So they're very slow about it. Many developments came out not at all as the developers had wanted at first, but smaller, more adapted to the area."

Local, democratic power is arguably the most fundamental and necessary requirement of sustainability; the two, as we have seen, almost always go together. Otto-Zimmermann says, "Globalists don't mind where they live, where they buy things, where those things come from. Borders, to them, just cause confusion, trouble, inefficiency. They are mentally torn down. But people who care about a certain place, they have real relations with it and with each other. This is *the fundamental disconnect* between the two approaches. And what we're learning now is that we really can have the best of both; internationalist populations that can see beyond local borders, and really care about what happens in other countries, but that also take care of their own homes, their *Heim*." The concept of local power, however, can also extend across entire nations, particularly in countries in Asia, Africa, and South America, where local power has for the last couple of centuries been usurped by colonial and then economic outsiders.

VIVA LA REVOLUCIÓN?

The implications of [the] Bagua [massacre] for Peru's democracy and South America's stability are ominous. The Bolivarian revolution has begun in Peru.
Caracas Gringo, right-leaning blogger[21]

Throughout the world the left tries to deceive the masses by presenting these governments [led by Chávez and others] as an alternative for the workers. This is clearly a typical vile lie of the bourgeoisie and their leftist servants aimed at subjecting the proletariat to the capitalist yoke.
Communist website[22]

The inflamed comments above from opposite ends of the political spectrum are typical media and Internet analyses of what has been happening in South America ever since Hugo Chávez was elected president of Venezuela ten years ago. That event was quickly followed by the elections of Luiz Inácio Lula da Silva in Brazil, Néstor Kirchner in Argentina, Tabaré Vázquez in Uruguay, Evo Morales in Bolivia, Michelle Bachelet in Chile, and Rafael Correa in Ecuador. This type of commentary is echoed in the mainstream media as well as obviously right- or left-biased sources. A more nuanced picture is starting to emerge, however, with the publication of more academic observations and studies. Share the World's Resources is a nonaffiliated, publicly supported NGO "with recommended consultative status at the Economic and Social Council of the United Nations." Anastasia Moloney, writing for their *World Politics Review*, says that, "While these leaders share some common characteristics, there are vast differences between them." Brazil and Chile represent the "soft" left, "modern, open-minded, reformist, and internationalist." Another left, more "nationalist, strident, and close-minded," she says, is exemplified by Chávez, Morales, and Correa.[23]

Moloney points out that all the newly elected leaders provide examples of "good old-fashioned populism," with their promises to redistribute money to the poor, common in the region since Juan Perón. They also exhibit the world's most serious resistance to neo-liberal globalization policies, typified by Brazil's treatment of the IMF, profiled below. And finally there's Chávez's brand of Bolivarism, which sees South America as an integrated region centered on social justice. Combined with these diverse elements are also expressions of indigenous values that she notes are coming especially out of Ecuador and Bolivia. In short, it's a heady and complicated mix, and, like a salad, has some sweet greens mixed in with the bitter.

What have South American politics got to do with individual, municipal, or national natural system sustainability? To start with, they obviously don't fit so easily into those old "left" and "right" boxes. South America is experimenting with a multiplicity of ways of governing and new ways of managing business and money. These experiments range from the mild to the radical, but they're worth watching, because the greatest danger the world faces, in both its natural and its social systems, is the recent human tendency to plant monocultures. Any single-system method of doing things, from agriculture to social and financial management, is flirting with death. Within your government or within your body, a lack of diversity means you're heading toward the flat line. Natural systems that are simple and contain only a few elements are the most vulnerable to the many changes that weather, disease, evolving pests, and microbes will throw at them. The same is true of social organizations, we are learning at our cost, as the one-economy-fits-all system of globalization begins to unravel. This inability to respond to change was one of the first principles the UNEP, the ILO, and the G8+5 noted following the 2008 economic meltdown.

One of the worst aspects of neoliberal economic globalization is its insistence on forcing virtually every government on Earth into the same economic mold. Once countries become fearful of being left out—even China fought to board the globalization money train and be governed by the WTO, IMF, and World Bank—the system can force its own small-government, big-business policies on each country. They typically encourage borrowing to expand economically and then demand "structural adjustment"—serious changes to social and political institutions, more private corporations, and less public input—as part of debt repayment plans. In the 1990s Malaysia stood up to the globalization monolith and refused to follow the adjustment prescriptions of the IMF. They were punished for their temerity and were thrown out of the club—and there they prospered. At that point Malaysia was almost the only country on Earth that had dared to defy the monocultured global economy, but today many South American countries have joined it. This has happened partly from the understandable South American desire to govern their local territories as they see fit, by nationalizing industries or holding on to various social programs. But it also has to do with a desire to manage finance, business, and money in ways that serve local interests and local values more directly. The methods these

Davids have used against the Goliath of global capital are, if nothing else, amusing to note.

After Néstor Kirchner was elected president of Argentina in 2003, he got away with defying the IMF because not much could happen to that country economically that hadn't already occurred when its economy collapsed in 2002—a catastrophe largely caused by globalization policies. What the IMF was demanding from this medium-sized country, richly endowed with both resources and an educated middle class, indicates that global bank organizations are far from being interested in money alone. They demanded that the Argentine economy show a *fiscal surplus of 4.5 percent of their GDP*, with ever-larger surpluses for the next few years, which was then, and today still is, virtually impossible. They wanted capital payments on its $15 billion debt, not just interest. And they wanted to help private utilities, water, sewage, and electric services make more money by demanding that Argentina set increases for their rates.

Journalist and political theorist Ken Hechtman has been watching South America's progression from IMF victim to rebel for some time. He notes that instead of acquiescing to their demands, Argentina gave the IMF only 3 percent increases to the GDP, no capital payments, and no utility increases. And then, a year later, "Kirchner suspended that agreement, fired his finance minister for wanting a new IMF deal, and only kept up with the payment schedule." He did this so that they could keep their export taxes, something Canada, a developed country but one based on resource exports, gave up long ago. A recent paper on the subject points out that "export taxes have been used by governments as a tool in their industrial policy and to raise revenue since the eleventh century. In fact it was the most important tool in industrial development while England was industrializing."[24]

"The IMF," Hechtman says, "always wants high sales and income taxes as part of Structural Adjustment Programmes, but they also always want all export taxes eliminated. That one fact is key to understanding what the IMF is really after." There's no big money in raw materials. Export taxes are a way for raw material exporters, like Canada and much of South America, to get a share of the value-added products, in order to offset the drain of their raw resources—turning logs into furniture, or bauxite into aluminum cans. These taxes don't wreck local economies because they put the tax burden on large industries outside the country that can afford to pay. Hechtman says, "If the IMF only cared about getting paid back, as they claim, they'd let their debtors keep this efficient and painless way of raising money. But they never do."

Hechtman has more to say about what happened in Chile in the first years of the twenty-first century, when that country owed the IMF money and Chilean leaders were told to privatize their banks. They refused; they also kept the capital flight controls that prevent foreign businesspeople from pulling their money out suddenly and destabilizing the country. "You gotta love the IMF," Hechtman says. "When all the banks failed in Chile in 2002, they wanted the Chilean and the international governments to bail them all out with billions and billions of taxpayer money. That wasn't socialism; that was 'restoring investor confidence.' But as soon as banks begin to make money again, they were back to wanting them privatized." In December 2005 Brazil really terrified the IMF by paying their $15 billion national debt back in full. The IMF tried to refuse it, once more proving the organization is focused on policy and power rather than money. Argentina did the same thing with their $10 billion debt, but they had to borrow the money from Venezuela to do it. "They're paying 9 percent interest to Chávez when they could have had 4.5 percent with the IMF," Hechtman notes; but Argentina's leaders feel it was worth it, "because there were no social or political policy strings attached."[25]

Helping fund some of this defiance, of course, is Venezuela's huge supply of that unsustainable commodity, oil. It has helped the continent's underdogs get elected without elite backing, it has helped them put in social programs, and it has helped them pay or defy their creditors. Back when oil prices were rocketing upward, Hechtman says, "the IMF was very concerned that the windfall not be spent on anything that might benefit Venezuelans. Instead, Chávez put the whole gain and then some into social development projects," winning yet more local, popular support. The most radical South American leader of all, the indigenous president of Bolivia, Evo Morales, allowed his last IMF agreement to expire in 2006, so the globalization giants cut him off from new loans. After all, against their explicit advice, he had renationalized the country's gas industry, which the IMF had forced a previous government to privatize.[26] These days, even though Bolivia is the poorest country in South America, Morales has used those gas royalties to increase health care accessibility to his people by 300 percent.

None of this is Communism; it's mostly a reordering of capitalism to make it more locally responsive and profitable for the people, which is one of the reasons Communist sources are even more scathing and critical of it than the right wing. It's closer to a soft socialism, similar to what Denmark, Sweden, or Holland enjoy, but it outrages global corporations and financiers because Latin American countries still have a huge resource base that the corporations were counting on as a cheap source for their future products and wealth. That's the fun story; but the part of the *revolución* that has South American countries taking national control of their resources doesn't bode entirely well for their natural systems or the global atmosphere and water cycles. The progressive (and overly optimistic) new constitutions of Ecuador and Bolivia aside, the whole region is spouting oil wells, highway and port projects, huge mining operations, gas drilling, and big dams as never before in history. Everywhere these megaprojects go in, there are protests by local and indigenous people. But unlike European and North American protests, theirs often end in violent death. The murders in 2009 of at least eighty native people involved in a blockade in Bagua, Peru, who were protesting the delivery of their Amazonian region into the hands of private developers, did prompt the Peruvian government to backtrack in its development legislation. But this government's core policies remain a constant threat to poor people who fear losing their lands and only means of support.

Evo Morales, the president of, Bolivia, is a native person who spent his childhood scavenging for food for his family. Morales had seven siblings, but four died in infancy, which is typical for indigenous families in this part of the world. As recently as the 1950s it was illegal for a native person to walk through the center of the capital, La Paz, or to go near the presidential palace or the cathedral; up to this day indigenous people in most Latin American countries have been starved, stolen from, killed by the millions, enslaved, or ignored. "They were (and are)," says Johann Hari in an article on Morales for *The Independent*, called "child-like... routinely compared to monkeys and apes."

Although his manner is shy and diffident, President Morales's policies reflect a deep desire for systemic change. As an indigenous person, he most likely understands that he does not represent himself but his entire group, and in their service, he has become, as Hari notes, "one of the most popular leaders in the democratic world... Millions of people are seeing doctors and schools for the first time in their lives."[27] Morales is using the nationalized gas royalties to do this, but something's happening that is far more radical than extracting finite resources and giving the money to the poor, rather than to the rich. Morales's very existence is detested by the white elites of eastern Bolivia, who up until now have always run the country. Yet he risked a dangerous, destabilizing referendum in order to pass a new constitution in 2008. It grants not only indigenous social and political rights but addresses core indigenous issues, which are nearly always about saving the land, taking less, and saving resources for the future and for their own sake.

The idea that land, animals, plants, rocks, and ecosystems also have rights to exist that must be legally respected has taken root in both the Ecuadorian and Bolivian constitutions. From the point of view of the planet's failing ecosystem, this is the real revolution in human thought emerging from South America today. It's important to remember that indigenous people and their worldview, although finally getting some attention from the rest of us, remain the least represented group in any position of prominence and power within any country or within international organizations.

This book has argued that the most likely reason for the continued isolation of humanity's oldest peoples is that the indigenous worldview sees humanity as *responsible for and deeply dependent upon natural systems*. That view is alien to nearly all other peoples of the world, who have been conditioned, both through religion and economics, to exploit natural systems almost without limit or respite. Despite the fact that indigenous people are marginalized within their territories and at decision-making political levels, they have somehow managed to survive into the twenty-first century, holding on to traditions that have, most of the time, treated their natural resources sustainably. So it is miraculous that even in a majority indigenous nation and one of the poorest and most viciously exploited places on Earth, an indigenous person has attained real political power. Support for Morales from other South American leaders may even mean that other forms of governance, beyond the current *de facto* oligarchies of wealth and influence, are starting to have more influence in the world. Certainly, the indigenous attitude toward the "rights" of our ecosystems is a step in the right direction.

THE VALUE OF MONEY

Money is like an iron ring we've put through our noses. We've forgotten that we designed it, and now it's leading us around.
Bernard Lietaer, Belgian currency expert[28]

One reason most of the cultures on this planet accept an economy based on managing resources for the short term instead of for the long term, as most traditional people have had to do, is that this kind of "investment" pays off better and much faster. The rewards the modern economy gives to people for using up the planet's natural resources as quickly as possible have increased exponentially in the past three hundred years. David Korten, author of *When Corporations Rule the World*, was part of that development process back in the 1950s and '60s. His experiences as an economic and business expert working for USAID and the Rockefeller and Ford foundations made him realize that most of us labor under a fundamental confusion about the very definition of what is valuable. "Wealth," he says, "is something that has real value in terms of meeting our needs and fulfilling our wants: the natural productive systems of the planet... and [physical things like] factories, homes, farms, stores, transportation and communications facilities... Modern money is only a number on a piece of paper or an electronic trace in a computer, that by social convention gives its holder *a claim* on real wealth. In our confusion we concentrate on the money, to the neglect of those things that actually sustain a good life."

In order to illustrate what he means, Korten suggests that we "think of a modern money economy as composed of two related subsystems. One creates wealth"—that is, it builds factories, stores, and farms; it employs people. It uses the extra production of nature in the form of plants, sunlight, or water at the same rate that it can be renewed. This can also be termed "living off nature's interest, while conserving its capital for future generations... This is how people have lived on the planet for many centuries," Korten says. "The other [subsystem] creates and distributes money as a convenient mechanism for allocating wealth. In a healthy economy, the money system serves as a dutiful *servant* of wealth creation, allocating real capital to productive investment, and rewarding those who do productive work in relation to their contribution."

Korten points out that money should never be the only or even the dominant medium of exchange. "One of the most important indicators of economic health is the presence of an active economy of affection and reciprocity, in which people do a great many useful things for one another with no expectation of financial gain." Anyone who has ever spent time in poorer countries, in rural areas, or in small towns knows exactly what this means. Korten adds, "Pathology enters the economic system when money, once convenient as a means of facilitating commerce, comes to define the life purpose of individuals and society." We can tell when a system has become pathological quite easily, he says. "When financial assets and transactions grow faster than growth in the output of real wealth, [that's] a strong indication that the global economy is getting sick."[29] That describes the 2008 economic collapse in a nutshell.

One of the financial experts that environmentalists have been consulting for many years, Bernard Lietaer, the author of *The Future of Money* who also helped design the euro, pointed out then that "Today's official monetary system has almost nothing to do with the real economy. Just to give you an idea, 1995 statistics indicate that the volume of currency exchanged on the global level is $1.3 trillion U.S. per day. This is *thirty times* more than the daily gross domestic product (GDP) of all the industrialized countries of the world put together."[30] Today it is many times that amount. This "money" being traded around the world is almost fully decoupled from any measurable physical wealth. Lietaer says, "Of that volume, only 2 to 3 percent has to do with real trade or investment; the remainder takes place in the speculative global cyber-casino. This means that the real economy has become relegated to a mere frosting on the speculative cake, an exact reversal of how it was just two decades ago." One reason this happened is that during the Nixon administration the United States decoupled paper money from gold, because they found pegging imaginary wealth to something physical was too limiting. Today a few countries still maintain a 10 percent reserve of deposits, which prevents the banking system from creating more than ten times as much currency; but as Korten says, "Today most money is created by borrowing. When a bank decides to grant a loan, they create that money out of nothing; but it still represents a claim on the real wealth of anyone who *wasn't* given a loan."

A given individual or corporate group can, under our current system, apply to a bank for a $2 billion loan for various kinds of business development. The bank grants the loan, which means they create that $2 billion out of thin air, and by social convention, agree that it is now in the hands of the corporation. So with that money, the corporation can go off to Peru, Saskatchewan, or Indiana and buy up mines, forests, factories—entire towns if they want—since they have the power of money against which the locals who used to have tenure over such things simply can't compete. This means, says Korten, that "increasingly, your money supply is being controlled by outsiders, by the banks. They're creating the money and they're taking the profits out of a country or place, essentially for renting that money to some group or corporation when they made out the loan." It's a system of pretend money. We make it up, then award huge amounts of it to certain groups, like investment banks or dot-com start-ups, but not to others, like disabled children or unemployed fishers. But there isn't any reason we can't use the pretend money any way we want. Only true wealth—water, soil, rocks, and calories stored in plants, animals, and organic materials like oil and gas—is limited by reality.

Besides being a founding member of the International Forum on Globalization, the main organization that fought economic globalization throughout the 1990s, David Korten and his wife, Frances, started a magazine called *Yes!* and an NGO, Positive Futures, that concentrates on trying to facilitate this important transition from one method of economic thinking to another. The 2008 economic meltdown gave Korten, like many other economic reformers, new energy. Working with partners like the Institute for Policy Studies, BALLE, and the New Economy Working Group, his most recent analysis is "Beyond the Bailout: Agenda for a New Economy" and provides recommendations for leaders to consider in designing the type of economy that can support human needs without destroying the planet on which future children depend. Although none of the world's powerful financial bodies—the WTO, the World Bank, the OECD, or the IMF—are embracing these ideas yet, they are not going to be able to ignore them for much longer. They run as follows:

1   "Market prices must internalize full social and environmental costs." That means that a toxic chemical in a fertilizer, for instance, must be costed out in terms of its effects throughout its life cycle. This action would reveal when such substances are too costly to use and provide the economic impetus for better technologies and products.

2   "Trade between nations must be in balance." That means that goods, chemicals, and subsidized foods cannot be dumped on poor countries without paying the same kind of dividends back for the commodities they produce. If they cannot produce an amount that would equal what they want to buy, that fact would preclude the long-term damage to society of going into debt.

3   "Investment must be local." These four words have tremendous effects and are a fundamental principle of sustainability, a primary criterion of the examples listed in this book.

4   "No player can be big enough to directly influence market price." Breaking up the monopolies and financial giants like General Motors or Citibank that can hold entire economic systems ransom because they are "too big to fail," used to be a fundamental prerogative of governments. Refusing to allow the situation to occur in the first place would take care of most of the entities seeking unsustainable bailouts.

5   "Economic power must be equitably distributed." This means allowing small local businesses, farms, and communities to have a say in what happens to their products and energies, and also letting small countries have control over how economies are structured. That way, minimum wages can reflect true need and progressive tax rates can control excesses in

income, contain health care costs, and regulate interest on
such things as mortgage and credit card interest rates.

6   "Every player must have complete information and there can
be no trade secrets—that is, no government-enforced intellectual property rights." This would take care of a great deal
of the more outrageously expensive medical drugs, as well as
destroy the lucrative patent motive for creating dangerous
new technologies like genetically engineered organisms or
nanotech.

7   Finally, and most importantly, "Markets must be regulated
to assure that these essential conditions can be maintained."
This requires "that we measure economic performance
against the results we really want"—healthy, prosperous societies, not a few outrageously rich individuals or corporations
surrounded by want.

One of the most important ways to structure a functional regulatory system involves replacing GDP, the Gross Domestic Product,
with GPI, a General Prosperity Index. GPIs measure health, education, and pollution levels—in other words, *quality* of life instead
of the number and amount of cash transactions undertaken every
year. The new, post-meltdown world is endowing the idea of these
new measurements, based on societal, not financial, values, with a
lot more power. What we use now, as Korten points out, the GDP,
"measures costs, not gains," which makes it clear why trying to
increase GDPs has led us into economic and natural resource crises.

Korten's most interesting and still radical suggestion is a proposal for what he calls "debt-free" money. Most people do not
realize that money is no longer issued by governments and that the
quantity issued is not based on some kind of wealth reserve, like

gold bars. It's not based on anything at all, in fact, except the desire of some individual or entity, which also enjoys power and connections, to have it. Since the 1980s almost all world currencies have been issued directly by private banks. As noted, when a bank agrees to make a loan for a business venture, the bank *creates* the dollars or yen or euros to finance that loan. The bookkeeping entry for that loan creates only the principal amount, not the interest that immediately begins to accrue, So the mere existence of this money will *automatically* take the bank's books into the red, unless the overall economy grows fast enough to generate more loans, to create more new money, so as to make interest payments on the old loans! This is palpably crazy, and no household could possibly be run this way; yet that's how Wall Street has set up the global economy.

What this creates is an immediate "demand for interest on nearly every dollar in circulation," which naturally puts the economy into default. Korten points out that both Thomas Jefferson and Benjamin Franklin, neither of them considered wild-eyed leftists, advocated replacing bank-created debt-money "with an alternative system in which the government creates *debt-free money* by spending it into existence" the same way we do now, but to fund public goods like health or education. This would not be in any way a more inflationary system than the one we already have; for example, where exactly is the interest on the $700 billion used to bail out companies after the 2008 meltdown coming from? More money creation, based on nothing except the fact that it's needed and governments and banks are in agreement about creating it.

The difference debt-free money would make is simply that the bookkeeping entry "would be made by government for a *public good*, instead of a private bank for *private profit*." Some of Barack Obama's and even Stephen Harper's economic rescue policies are a *de facto* move toward government-created, public-good money, especially if they create more jobs in service industries, like teaching and nursing. Money created for infrastructure construction to provide jobs, as in Franklin Roosevelt's New Deal, is the same kind of thing. This money is being created by the government and would be intended to serve the public good by preventing job losses and recession. It would be very easy to take the money that was created to rescue failing private corporations like AIG or General Motors, and instead invest it in new technologies like solar power, organic farms, or public transport.

People are beginning to realize that a major reason for the huge pressures on our finite and dwindling true wealth of forests, farms, fresh water, and seashores is that old-fashioned debt-money absolutely *requires infinite growth* in the economy, on a finite planet, mind you, just to pay the interest that was created along with it. Debt-free money would not require the growth-for-growth's-sake development that currently results in more housing, more fishing, and more farm production—that uses up long-term national resources in order to feed the maw of accruing interest. Without the urgent demand to pay interest, development could not only proceed at a milder and more sustainable pace, but, as in Rajasthan and Freiburg, *time* could be spent to determine its ultimate utility to the people in the area.

At the moment only private desires for more wealth are being served when money is issued to an individual or large company intent on, for example, taking a gamble that destroying more coastline with hotels, roads, or shrimp farms will make a quick fortune. If we issued money to publicly answerable entities to build new schools, educate populations about family planning, or fund medical care, development that supports positive human endeavors would not stop, as critics cry, but would obviously increase. The dividends of that development would benefit everyone instead of the wealthy few, and more public jobs would replace some of those private jobs—not such a big deal when we consider that for the past twenty years, we've legislated the opposite.

In other words, we used to use money like that as recently as the 1960s, and our finances and futures were a lot more stable. Such fiscal policies would quickly get us more orderlies and nurses working in local hospitals, more construction workers employed on needed infrastructure like railroads or bridge repair, and more farmers taking care of their animals humanely. This lending would also automatically be subject to public oversight, as opposed to private manipulations set up to respond uniquely to the constant interest pressures of debt-money. Even if it does sound too idealistic to imagine, considering that the system we're now working under will, without the slightest doubt, eventually destroy this planet and wipe out our species, some serious rethinking of how we run our economies is surely worth a try.

Fortunately, elements of Korten's alternative system are being introduced in small, medium, and even fairly large ways, all over the world. In the first edition of this book, we talked about the local currency movement, where cities and neighborhoods print up their own forms of exchange to keep money within the area, a movement that continues to gain adherents. Revolutionary banks, like ShoreBank in Chicago, Triodos Bank in the Netherlands, and Germany's GLS Gemeinschaftsbank, are thriving and expanding, even if we still don't have nearly enough of them. These institutions lend money to organic or socially responsible start-ups, finance international microcredit schemes, and help alternative business systems like cooperatives.

With more than a billion dollars in assets, Triodos and GLS aren't exactly small, but along with smaller, community-based banks and credit unions, all of these alternative banking systems, without exception, have been able to laugh at the financial crisis that has destroyed or tainted major financial institutions, from Fannie Mae to Wells Fargo and the World Bank. That's because they haven't been indulging in the speculative "casino" economy but manage money in the same ways that householders know will keep them in the black.

Triodos money helped set up the Essential Trading Co-operative, one of the largest worker-owned co-ops in Britain, which imports fair-trade goods and wholesales organic foods to six hundred retailers around the U.K. Triodos Belgium provided a large loan for the construction of a huge eco-building that houses the Oxfam World Shops in Ghent. There are many other bank services available, such as their most brilliant venture, Triodos Match, Ltd., which acts as a matchmaker to pair up social and environmental businesses that need capital with people who have capital—and, frequently, relevant skills and experience as well. These investment "marriages" typically work for businesses trying to raise between £20,000 and £500,000 (US$28,000 and US$700,000). Besides all this, Triodos offers normal banking services and is active in outreach projects such as forgiving Third World debt and supporting organizations like the Environmental Law Foundation. They also give a lot of their profits to charities. They're definitely not the kind of bank we're used to.[31]

Canada continues to lag behind in creating such wonders, having only one such bank, VanCity in Vancouver; but there are two similar financial institutions in the U.S. Chicago's ShoreBank was founded in 1973 mainly to help entrepreneurs in inner-city neighborhoods get the start-up funding they need to take control of their lives. So far they've sent out $600 million in loans to help revitalize the lives of thousands of families and businesses in the tough south and west sides of Chicago. That completely social mandate expanded with the founding of their affiliate, ShoreBank Pacific, opened, not surprisingly, in Portland, Oregon. Its focus is environmentally sustainable and community development projects.

The institutions of the credit union movement (called *caisses populaires* in Quebec) have been around for close to a hundred years now. Many have become far too much like conventional banks, but, because they essentially operate as co-ops, they could be a foundation for ethical, sustainable banking models if their members wanted them to be. Financial reform movements like William Spademan's "Common Good" banks in Maine are looking to expand and increase the usefulness of such models. Very small tweaks in legislation could protect small community banks like Home Federal, in Sioux Falls, South Dakota, or the First National Bank of Orwell, Vermont, from shareholder pressure to take risky loans or to engage in dangerous speculation, as well as from buyouts and hostile takeovers. Presently both those examples and thousands more have weathered the dot-com, housing, and financial bubble bursts with perfect equanimity. They have always functioned like the *It's a Wonderful Life* type of small bank or credit union, run by local people, offering modest returns; not as gambling casinos controlled by insanely greedy risk-takers a world away.

Spademan's idea for "Common Good" banks would lock speculators out more formally, by writing founding documents requiring them to be depositors, with bylaws that allow each shareholder one vote, as opposed to granting one vote per share of stock. This is what credit unions do to keep their membership, instead of the money, empowered. But they're barred from turning a profit. Common Good banks could be profitable, but their profits would have to be spent on charities within their local community; that is, like many cooperatives, they'd have to work hard in order to share. The real profit would be the dramatic increase in a bank's role in the community, as well as a steadily more viable community in which to work. Spademan's ideas are almost as radical as Korten's and are well on their way to becoming reality. In spring 2009 he hired twenty-three bank division coordinators in twenty-three different communities, to gather the new banks' financing.

TAKING VISIONARY ACTION

There is no more exhilarating and realized life, I believe, than a life of service.

Jarid Manos[32]

We're at a very serious crossroads in our evolution and can no longer say the choices before us aren't clear. If human beings continue to value our abstract, social fantasy of money and expect it to grow endlessly, we will destroy everything that makes life not only worth living, but possible, from drinkable water and green countrysides to blue-ringed octopi and food that doesn't poison our babies. We face having to redefine what wealth means; and that also entails redefining the whole idea of human happiness.

People have always wanted to be prosperous because we have assumed that wealth makes us happy. Studies that look at what really does make us happy have proliferated in recent years, and the results of many suggest otherwise—that once the basic needs of food, housing, and clothing are met, increasing amounts of income do not mean increasing amounts of happiness. What seems to affect happiness levels most is having useful jobs, good health, and a feeling of purpose and connection with the world. In fact giving money away to others has recently been found to provide one of the biggest happiness boosts imaginable.[33] All of which suggests that we are social animals far more than we are money animals.

Proof of our social nature can be found in the need to aggressively market products and push consumption. If we really did want all the products that come out every year, we'd find out about them and acquire them on our own; manufacturers could save on those huge marketing expenses. No one has to be pressured to buy food, basic tools, and warm clothing. But companies do have to advertise designer clothes and the latest electronic gadget, because most people don't want these things initially unless they are encouraged to do so, most often for purely social reasons.

Luckily, most people are still able recognize true needs over artificially induced ones. Between 1996 and early 2008, dishwashers and microwaves were classified as "necessities" by 40 percent and 70 percent, respectively, of the U.S. population; when the bubble economy of those years burst, these appliances were viewed as necessities by a mere 15 percent and 20 percent of the population.[34] That shows how quickly people can adjust their ideas of what they "need" to be happy, to what people around them are doing or saying. That's a helpful human ability we need to perfect very quickly, because we can't adjust humanity's true needs the same way. We really can't do without clean water, food, and air.

One reason the idea of an ever-growing economy has been so attractive worldwide is that it permits us to have an ever-growing population of humans. Every species on Earth is hard-wired to increase its numbers. That's usually difficult to do, given the constraints of habitat, food, and water availability. All species also have to deal with predators, accidents, and diseases. The way nature is set up, overproduction of offspring provides food for other species, and adult levels of a species remain fairly stable. But we humans have removed so many predators, from grizzly bears to smallpox, that we've been able to increase our numbers to the point that we're eroding all of our available habitat, including future generations' supply of food, land, and water. If we want to stay on this finite planet—and there are no spaceships leaving for another one any time soon—we have to embrace the idea of a *finite population*.

Granting social and educational status to women all over the world and providing them and their young children with medical care will get human numbers back down to where they need to be in little more than a generation. In short, the old kindergarten "sharing" rule plays a big part in attaining what we want—and what we need as well. We have to get very busy on the kind of social pressures and rewards that make controlling our numbers easier so that the future can become less bleak than it's looking right now.

If you want to imagine a really rosy eco-future, check out the last chapter of Margaret Atwood's recent book *Payback*. In Atwood's retelling of Charles Dickens's *Christmas Carol*, Ebenezer Scrooge is a financier and CEO; the ghost of the present "looks a little like David Suzuki, and a little like Al Gore"; and the ghost of the future is a "glinty-eyed" futures trader who takes Scrooge to a city where the people are wearing "natural-fibre clothing and riding on bicycles... using power from wave-generation machines and from solar installations on the tops and sides of their buildings." Their diets are organic, and, like Cubans, they grow food on their former front lawns. In this world, "evil bottom-scraping fishing practices have been abandoned... all religious leaders have realized that their mandate includes helping to preserve the Almighty's gift of the Earth and have condoned birth control." Even global warming "has been dealt with at a summit during which world leaders gave up paranoia, envy, rivalry, power-hunger, greed... and got on with it."

The spirit of Christmas future, which sometimes manifests as a cockroach—since in some futures there are no humans—shows Scrooge an alternative future based on business as usual, going on just the way we are now—a future that is a nightmare. When Atwood asks herself if the first vision will triumph, if people will live more like the inhabitants of Freiburg, who have learned "to calculate the real costs of how we've been living, and of the natural resources we've been taking out of the biosphere," she answers, "My best offer is Maybe."

Like Margaret Atwood, we've tried to write an honest book. We know that the criteria to judge the solutions we've outlined will enable people to recognize the sustainable ones. The real eco-solutions that fit in with the five criteria of sustainability outlined in Chapter 1—biomimicry, flexibility, humility, bottom-up democracy, and lofty goals—are not based on gadgets, markets, fads, or theory. They are not based on the assumption that we can have our cake and eat it too; that is, that we can conserve ecosystems while having huge numbers of babies, enjoying ever more lavish lifestyles, and continuing to use existing financial, political, and economic paradigms. The solutions in this book aren't like that. They will cost us something.

To begin with, we'll have to develop a whole new value system, far more in tune with the restraint and true "conservation" practiced by traditional peoples. The fact that in the long run our new value system will probably make us happier doesn't mean there won't be a certain amount—in fact, probably a lot—of irritation in some quarters as well as some privation in others. Very few of us will be flying around in airplanes, riding on gas-powered lawn mowers, or basking in front of air conditioners in this more realistic vision of the future. We will, however, have healthier bodies, swimmable rivers, more delicious food, more beautiful forests, and even some fish left in the sea. Our children will have something to look forward to. Hopefully, that exchange of benefits will make sense to most of us.

In this book we haven't tried to figure out if these solutions are politically possible, or if the simple rules of sustainability we keep referring to can overcome natural inertia or relentless opposition from very powerful interests, who will defend the old industries and financial systems very bitterly. There are thousands of experts in academia and government and community organizations working on the details of how to achieve and balance political expediency and social acceptance. In this book, on the contrary, we have simply tried to assess as scientifically as possible *what natural systems need to survive*, not what human beings would like or are willing to contribute to their survival. We think it's becoming obvious that the ethic of restraint is one that human beings no longer have much choice about adopting. Either people will learn to contribute to ecosystem survival, or our species will disappear. We will also learn that if science proves one thing with utter clarity, it's that human beings are not in the driver's seat on this planet, however much we might like to think we are. People depend on and are literally in debt to the natural systems that keep trying to support us.

Which brings us back to Margaret Atwood. Her book is about debt; it's about borrowing and then paying back whatever it is that people value and think they need. At the end of *Payback*, the reformed Scrooge wakes up as in the old Dickens story, goes to the window and finds the world very beautiful and newly "fragile, like a reflection on water." Like most of the people living in the industrialized world, Atwood's version of the money-obsessed Scrooge has several cars and enjoys fancy vacations, nice clothes, and fine wines.

He's gotten rich taking what he wants—as well as what he needs—away from the planet. And now he realizes that he doesn't really own anything, not even his body. "Everything I have is only borrowed. I'm not really rich at all, I'm heavily in debt," he cries at the end. "How do I even begin to pay back what I owe? Where should I start?"

Jarid Manos, quoted at the beginning of this section, founded the Great Plains Restoration Council. It puts damaged children from inner cities and native reserves to work restoring the prairie outside of Fort Worth, Texas. We all might consider following his advice and do ourselves the inestimable favor of offering a life of service to the Earth. Our individual powers, including our power to work with each other, as so many people in this book have demonstrated, are vastly beyond what we usually think they are. Anyone can become a Maude Barlow or a Rajendra Singh, or just as usefully, get out there and help them! We need to remember that every effort people make, however small, and every fruit we pick, however low to the ground, takes us another step toward a real future on our very real planet. Today the Earth's basic functions, cycles, and systems, from her oceans and forests to her rivers and grasslands, from the carbon and water cycles to the atmosphere, have come to need the actions of humans, absolutely as much as we need theirs. The revolution of the twenty-first century is that the human–Earth relationship has become totally reciprocal. The endnotes in this book can guide readers to the hundreds of organizations working to achieve sustainability. They are all hoping to attract more help, and there's very little time left. Every one of us needs to become an agent of change for the Earth.

NOTES

CHAPTER 1: VIVA LA REVOLUCIÓN

1   Paul Hawken. *Blessed Unrest.* New York: Penguin Books, 2007. p. 186.

2   Linda Gyulai. "Walking the green walk." *Montreal Gazette,*
    Nov 29, 2008. pp. 3–4 (B1).

3   Carol Estes. "Living large in a tiny house." *Yes! Magazine,* Winter 2009.
    www.yesmagazine.org/issues/sustainable-happiness/living-large-
    in-a-tiny-house

4   Paul Hawken with Amory Lovins and L. Hunter Lovins.
    Natural Capitalism: Creating the Next Industrial Revolution.
    New York: Little, Brown & Co., 1999. p. 9.

5   Bill McDonough. "How do you love all the children?" *Yes! Magazine,* Fall 1999.
    www.yesmagazine.org/issues/power-of-one/how-do-you-love-all-the-children

6   William McDonough and Michael Braungart. "The NEXT industrial
    revolution." *The Atlantic,* Oct 1998. pp. 82–92.

7   Sarah van Gelder. "Corporate futures." Interview with David Korten and
    Paul Hawken. *Yes! Magazine,* Summer 1999. www.yesmagazine.org/issues/
    cities-of-exuberance/752

8   Ibid.

9   Ibid.

10 Interview by Holly Dressel, 2001.

11 John Mackay. "Canada: Listeriosis epidemic continues—Liberal and Tory
    policies culpable." World Socialist Web Site, Oct 14, 2008.
    www.wsws.org/articles/2008/oct2008/list-014.shtml

12 See I. van Loo, et al. "Emergence of methicillin-resistant *Staphylococcus
    aureus* of animal origin in humans." *Emerging Infectious Diseases,*
    Vol. 13, Number 12, Dec 2007. pp. 1834–39.

13 Miguel Altieri and Peter Rosset. "Ten reasons why biotechnology will not
    ensure food security, protect the environment and reduce poverty in the
    developing world." *Food First,* Mar 7, 2005. www.foodfirst.org/en/node/305/;

Lori Ann Thrupp, ed. *New Partnerships for Sustainable Agriculture.*
Washington, DC: World Resources Institute, 1997; UNDP. *Agroecology:
Creating the Synergism for a Sustainable Agriculture.*
New York: United Nations Development Programme, 1995.

14　Amanda Griscom. "RE: William McDonough." *FEED,* June 12, 2000.

15　Ibid.

16　Interview for The Suzuki Diaries documentary series. *The Nature of Things.*
CBC Television, Aug 2008.

17　Chris Turner. "The wind at his back." *The Globe and Mail,* Aug 2, 2008.

18　Ibid.

19　Interview for The Suzuki Diaries documentary series, Aug 2008.

20　"The gospel of green." Interview by Bob McKeown. *The Fifth Estate,*
CBC Television, Nov 2008.

21　International Institute for Democracy and Electoral Assistance figures for
voter turnout are available at www.idea.int/vt

22　Quoted in Stephen Leahy, "Future prosperity is in green technologies."
Straight Goods, Oct 30, 2008.

23　Mark Schapiro. "Let's go Europe." *Mother Jones,* Nov/Dec 2008. pp. 69–71.

24　See "The economics of ecosystems and biodiversity," *EUROPA Environment,*
www.ec.europa.eu/environment/nature/biodiversity/economics

25　Quotes from Solheim and Sukhdev in, "Global green New Deal: Environmentally focused investment historic opportunity for 21st-century prosperity
and job generation." United Nations Environment Programme press release,
Oct 22, 2008. www.unep.org/Documents.Multilingual/Default.asp?
DocumentID=548&ArticleID=5957&l=en

26　John Bellamy Foster. *The Ecological Revolution: Making Peace with the Planet.*
New York: Monthly Review Press, 2009. p. 105.

27　Govinda R. Timilsina, Nicole LeBlanc, and Thorn Walden.
Economic Impacts of Alberta's Oil Sands, Vol. I. Calgary: Canadian Energy
Research Institute, Oct 2005. p. 13–15. www.ceri.ca/Publications/documents/
OilSandsReport-Final.pdf

28　"U.S. Congress upholds restrictions on high-carbon fuels." CBC News,
Sept 25, 2008. www.cbc.ca/world/story/2008/09/25/tarsands-congress.html

29　Diana Gibson. "When does the ship come in for the average Albertan?"
Union, Winter 2008. www.afl.org/upload/unionwinter2008.pdf

30 Doug Saunders. "Frugal Norway saves for life after the boom."
    The Globe and Mail, Jan 31, 2008. www.theglobeandmail.com/archives/
    frugal-norway-saves-for-life-after-the-boom/article664155/

31 Ibid.

32 Margaret Munro. "Biofuel: salvation or perdition?" Canwest News Service/
    Montreal Gazette, May 3, 2008.

33 Quoted in Edith M. Lederer, "UN expert calls biofuel 'crime against humanity.'"
    Associated Press, posted on LiveScience.com, Oct 27, 2007.
    www.livescience.com/environment/071027-ap-biofuel-crime.html

34 See R. Samson, et al., *Analysing Ontario Biofuel Options: Greenhouse Gas
    Mitigation Efficiency and Costs.* BIOCAP Foundation of Canada, 2008.
    www.biocap.ca/reports/BIOCAP-REAP_bioenergy_policy_incentives.pdf

35 George Monbiot. "An agricultural crime against humanity: Biofuels could kill
    more people than the Iraq war." The *Guardian,* Nov 6, 2007.

36 C. Ford Runge and Benjamin Senauer.
    "How biofuels could starve the poor." *Foreign Affairs,* May/June 2007.
    www.foreignaffairs.org/20070501faessay86305c-ford-runge

37 Rhett A. Butler. "U.S. biofuels policy drives deforestation in Indonesia,
    the Amazon." *Mongabay.com,* Jan 17, 2008.
    http://news.mongabay.com/2008/0117-biofuels.html

38 Andrew Bounds. "OECD warns against biofuels subsidies."
    Financial Times, Sept 10, 2007.

39 Roger Samson, et al. *Analyzing Biofuel Options: Greenhouse Gas
    Mitigation Efficiency and Costs.* Brief by REAP Canada, submitted to the
    House of Commons, Standing Committee on Agriculture and Agri-Food
    on amendments to *Bill C-33, Act to Amend the Canadian Environmental
    Protection Act,* Feb 27, 2008. www.reap-canada.com/library/Bioenergy/AAFC_
    Standing_Committee_Briefing.pdf

40 Ibid.

41 Interview by Holly Dressel, Fall 2009.

42 See www.nacosar-canep.ca/faqs_en.php and www.cosepac.gc.ca/eng/sct4/
    index_e.cfm

43 Raúl Zibechi. *Ecuador: The Battle for Natural Resources Deepens.*
    Americas Program Report. Washington, DC: Center for International Policy
    (CIP), Oct 26, 2009. http://americas.irc-online.org/am/6521#_ftn12

44 Bruno Latour. *Politics of Nature*. Cambridge, MA: Harvard University Press, 2004, p. 37. Emphasis added.

45 Cormac Cullinan. "If nature had rights." *Orion Magazine*, Jan/Feb 2008. www.orionmagazine.org/index.php/articles/article/500

46 Raul Burbano. "Refounding Bolivia: Struggle for the new constitution." *Green Left Weekly*, Oct 17, 2008. www.greenleft.org.au/2008/771/39748

47 Cullinan.

CHAPTER 2: USING COYOTES TO GROW GRASS

1 Mario Osava and Alejandro Kirk. "Quest for the Amazon turtles." Inter Press Service News Agency, Mar 6, 2009. www.ipsnews.net/print.asp?idnews=46013

2 Susan Milius. "Surprising number of creatures bipolar." *ScienceNews*, Vol. 175, Number 6, Mar 14, 2009. www.sciencenews.org/view/generic/id/40927/title/Marine_census_Surprising_number_of_creatures_bipolar; see ATV interview with Uzma Kahn at www.youtube.com/watch?v=pw5DwGdx238

3 Osava and Kirk.

4 William Cronon. *Changes in the Land*. New York: Hill and Wang, 1983. pp. 22 and 107.

5 "A political history of the Adirondack Park and Forest Preserve." Articles in the "History" section of www.adirondack-park.net.

6 Ibid.

7 Ibid.

8 Verplanck Colvin. *Annual Report on the New York State Museum of Natural History*, 1871. Quoted in "Political history of the Adirondack Park and Forest Preserve: Prior to 1894 constitutional convention" at www.adirondack-park.net.

9 Philip G. Terrie. *Contested Terrain: A New History of Nature and People in the Adirondacks*. Syracuse, NY: Syracuse University Press, 1997. p. 106.

10 See Philip G. Terrie, "Behind the blue line." *Adirondack Life*, Jan/Feb 1992; Philip G. Terrie. "The Adirondack paradigm." *Adirondack Life*, Collector's Issue, 1999; Paul Schneider. *The Adirondacks: A History of America's First Wilderness*. New York: Henry Holt & Co., 1997.

11 See http://nationalzoo.si.edu/Education/ConservationCentral/challenge/textsummary.cfm

12 Interview by Holly Dressel, 2001.

13 *Government of India Tiger Task Force Report,* Section 3.4.
New Delhi: Innovative Protection Agenda, Mar 17, 2005. pp. 63–69.

14 Kevin Kelly. "Assembling complexity." In *Out of Control.*
New York: Addison-Wesley, 1994. p. 68.

15 Ibid.

16 Allan Savory with Jody Butterfield. *Holistic Management: A New Framework for Decision Making.* Washington, DC: Island Press, 1999, p. 37.

17 Ibid, pp. 195–215.

18 Kenneth Brower. "Leopold's gift." *Sierra Magazine,* Jan/Feb 2001.
www.sierraclub.org/sierra/200101/leopold.asp

19 Jack Southworth. "Grazewell." *Managing Wholes.*
www.managingwholes.com/graze-well.htm

20 Interview by Holly Dressel, 2001.

21 Columbia River Inter-Tribal Fish Commission. *Bringing Salmon Back to the Columbia River: How Native American Tribes Are Implementing a Watershed-Wide Plan.* Portland, OR: Columbia River Inter-Tribal Fish Commission, Oct 2004. www.leadershipforchange.org/insights/research/files/31.pdf

22 John Ralston Saul. *A Fair Country: Telling Truths about Canada.*
Toronto: Penguin, 2008. p. 82.

23 Government of Canada. *2006 Census Bulletin: Data on Aboriginal Peoples.*
www.metrovancouver.org/about/publications/Publications/Aboriginal AffairsBulletin12.pdf

24 Fikret Berkes. *Sacred Ecology.* New York: Taylor & Francis, 1999, p. 126.

25 Interview by Holly Dressel, 2009.

26 W. Pearson and R.D. Andrews. "Can the U.S. National Park model be applied successfully to a unique and culturally distinct society? A case study of the Maasai and Amboseli National Park." July 2, 2002. [Page reference?]

27 Quoted by John Ralston Saul in *A Fair Country.* p. 64.

28 UNEP. *People, Parks and Wildlife: Guidelines for Public Participation in Wildlife Conservation.* Nairobi: UNEP publishing, 1988. p. 73.

29 Pearson and Andrews. p. 13.

30 J.A. McNeely and K. Miller, eds. *National Parks, Conservation, and Development.* Washington, DC : Smithsonian Institution Press, 1984. p. 96.

31 M. Wells and K. Brandon. *People and Parks: Linking Protected Area Management with Local Communities.* Washington, DC: The World Bank, 1992. p. 70.

32 Pearson and Andrews. p. 14.

33 K.P. Ghimire and M. Pimbert, eds. *Social Change and Conservation: Environmental Politics and Impacts of National Parks and Protected Areas.* London: Earthscan Publications Limited, 1997. p. 115.

34 Amy Corbin. "Haida Gwaii, British Columbia," *Sacred Land Film Project.* www.sacredland.org/world_sites_pages/Haida_Gwaii.html

35 Kim Petersen. "The struggle for Haida Gwaii." The *Dominion,* Issue 23, Nov 6, 2004. www.dominionpaper.ca/original_peoples/2004/11/06/the_strugg.html

36 Erin Millar. "Trophy lives." *BC Business,* Sept 3, 2008. pp. 5–6. www.bcbusinessonline.ca/bcb/top-stories/2008/09/03/trophy-lives

37 Corbin.

38 James K. Agee and Darryll R. Johnson, eds. *Ecosystem Management for Parks and Wilderness.* Seattle: University of Washington Press, 1988. p. 226. Emphasis added.

39 Quoted in Amanda Griscom, "RE: William McDonough." *FEED,* June 12, 2000.

40 Quoted in John Brazner, "Legislating sustainability." *Alternatives Journal,* Vol. 34, Number 4, May 26, 2009. pp. 17–18.

41 Janet Larkman. "Presentation to the panel hearing for the Whites Point Quarry and Marine Terminal Project." June 28, 2007. www.ceaa.gc.ca/010/0001/0001/0023/001/WP-1784-047.pdf

42 "Rejecting Digby Neck quarry: NS Environment Minister Parent turning the tide in Nova Scotia." Sierra Club of Canada press release, Nov 21, 2007. www.sierraclub.ca/atlantic/programs/economies/digbyquarry/press.htm

CHAPTER 3: AVOIDING VENUS

1   Quoted in Michelle Mech. "Briefing paper for Canadian government on ecological and economic urgencies of global warming." *West Coast Climate Equity,* Dec 8, 2008. http://westcoastclimateequity.org/?p=1519&print=1

2   That is, from 280 ppm to 360 ppm.

3   Andrew Weaver. *Vancouver Sun,* Feb 8, 2001.

4   L. Radovic. "Fossil fuels: Environmental effects." June 12, 1995. www.ems.psu.edu/~radovic/env_fossil.html

5   "Energy: Harvesting the wind." PBS television series e^2, Episode 6. Video clip available at www.pbs.org/e2/episodes/201_harvesting_the_wind_trailer.html

6   *Select Committee on Science and Technology Fourth Report.*
www.publications.parliament.uk/pa/cm200304/cmselect/
cmsctech/316/31602.htm

7   Cristina L. Archer and Mark Z. Jacobson. "Evaluation of global wind power."
Journal of Geophysical Research—Atmospheres, Vol. 110, 2005.
www.stanford.edu/group/efmh/winds/global_winds.html

8   Lester R. Brown. "Want a better way to power your car? It's a breeze."
The *Washington Post,* Aug 29, 2008. www.washingtonpost.com/wp-dyn/
content/article/2008/08/29/A R 2008082902334.html; David Bradley.
A Great Potential: The Great Lakes as a Regional Renewable Energy Source.
Buffalo, N Y: Green Gold Development Corporation, Feb 6, 2004.
http://greengold.org/wind/documents/107.pdf; Adrian Lema and
Kristian Ruby. "Between fragmented authoritarianism and policy coordina-
tion: Creating a Chinese market for wind energy." *Energy Policy,* Vol. 35, Issue 7,
July 2007; Jonathan Watts. "Energy in China: 'We call it the Three Gorges of the
sky. The dam there taps water, we tap wind.'" The *Guardian,* July 25, 2008.

9   Elizabeth May and Zoë Caron. *Global Warming for Dummies.*
Mississauga: John Wiley & Sons, 2009. pp. 216–17.

10   Quoted in "M I T-led panel backs 'heat mining' as key U.S. energy source."
M I T news, Jan 22, 2007. http://web.mit.edu/newsoffice/2007/geothermal.html

11   Jenny Haworth. "If Portugal can rule the waves, why not Scotland?"
The *Scotsman,* Sept 24, 2008. http://news.scotsman.com/opinion/If-Portugal-
can-rule-the.4520629.jp

12   "C E T O overview." On Carnegie Wave Energy website.
www.carnegiecorp.com.au/index.php?url=/ceto/ceto-overview;
John Brooke, ed. *Wave Energy Conversion.* Oxford: Elsevier, 2003. p. 7.

13   May and Caron, p. 214.

14   Michael Schirber. "Whatever happened to geothermal energy?" *Live Science,*
Dec 4, 2007. www.livescience.com/environment/071204-geothermal-energy.
html; "New Zealand geothermal fields." On the New Zealand Geothermal
Association website. www.nzgeothermal.org.nz/nz_geo_fields.html

15   May and Caron, p. 215; U.S. Department of Energy/M I T Panel.
The Future of Geothermal Energy. Idaho Falls: Massachusetts Institute
of Technology/Idaho National Laboratory, 2006. http://geothermal.inel.gov/
publications/future_of_geothermal_energy.pdf

16 See "Geothermal power," *Science Clarified.* www.scienceclarified.com/scitech/
    Energy-Alternatives/Geothermal-Power.html

17 Ibid.

18 Ibid.

19 Quoted in E.A. Wallis Budge, *The Dwellers on the Nile: The Life, History,
    Religion and Literature of the Ancient Egyptians.* New York: Dover Publications,
    1977 (first published 1926). pp. 250–53.

20 Craig Whitlock. "Cloudy Germany a powerhouse in solar energy."
    Washington Post Foreign Service. May 5, 2007. www.washingtonpost.com/
    wp-dyn/content/article/2007/05/04/AR2007050402466_pf.html

21 Manny Diaz, speaking at the Miami Environment and Energy Forum,
    Oct 13, 2008. www.cfecoalition.org/PDFs/Mayor_Diaz_Speech.pdf

22 Interview with Amory Lovins. "Let the little guys play." *Mother Jones,*
    May/June 2008, p. 45.

23 "Chicago outlines plan to slash greenhouse gases." *Chicago Tribune,*
    Sept 18, 2008. www.chicagobreakingnews.com/2008/09/chicago-outlines-
    plan-to-slash-greenhouse-gases.html

24 "U.S. mayors resolve to avoid burning tar sands oil." Environment News
    Service, June 28, 2008. www.ens-newswire.com/ens/jun2008/
    2008-06-28-01.html

25 "c40 goals for UNFCCC." c40 Cities website.
    www.c40cities.org/about/goals.jsp

26 Editorial. *Ottawa Citizen,* February 19, 1999.

27 "The OECD Environmental Outlook identifies 'red lights' for the future."
    March 29, 2001. www.oecd.org/document/50/0,3343,en_2649_34305_
    2345906_1_1_1_1,00.html

28 Hanno Beck, Brian Dunkiel, and Gawain Kripke. *Citizens' Guide to
    Environmental Tax Shifting.* Washington, DC: Friends of the Earth, June 1998.

29 May and Caron, pp. 146–47.

30 See "What is a carbon offset?" www.davidsuzuki.org/Climate_Change/
    What_You_Can_Do/carbon_offsets.asp

31 Kevin Drum. "10 ways to trade up." *Mother Jones,* Mar/Apr 2009, pp. 49–53.

32 Tyler Hamilton, "Americans seeing the light on carbon tax." *Toronto Star,*
    Jan 12, 2009. www.thestar.com/business/article/569202

33 Pew Center on Global Climate Change. "Cap and trade." *Climate Change 101.* Arlington, VA: Pew Center on Global Climate Change and the Pew Center on the States. www.pewclimate.org/docUploads/Climate101-CapTrade-Jan09.pdf

34 Hamilton.

35 Local worker for Grameen Shakti, Mawna unit, selling one-house solar lighting systems on credit. Quoted on "Energy: Energy for a developing world." PBS television series e^2, Episode 2. Video clip available at www.pbs.org/e2/episodes/202_energy_developing_world_trailer.html

36 James J. MacKenzie, Roger C. Dower, and Donald D.T. Chen. *The Going Rate: What It Really Costs to Drive.* Washington, DC: World Resources Institute, 1992. p. 5. pdf.wri.org/goingrate_bw.pdf

37 "Energy: Energy for a developing world." PBS television series e^2, Episode 2.

38 Terry Pratchett. *The Fifth Elephant.* New York: HarperTorch, 2000. p. 58.

39 Paul Roberts. "Seven myths of energy independence." *Mother Jones,* May/June 2008. p. 37.

40 Emily Rochon, et al. *False Hope: Why Carbon Capture and Storage Won't Save the Climate.* Amsterdam: Greenpeace International, 2008. p. 5. www.greenpeace.org/raw/content/usa/press-center/reports4/false-hope-why-carbon-capture.pdf

41 David Fleming. "Nuclear energy in brief." In *The Lean Guide to Nuclear Energy: A Life-Cycle in Trouble.* London: The Lean Economy Connection, 2007. www.theleaneconomyconnection.net/nuclear/Nuclear.pdf

42 Daniela Estrada. "Chile: Biofuels head to the forests." *Tierramérica*/Inter Press Service News Agency, Feb 6, 2009. ipsnews.net/news.asp?idnews=45702. Emphasis added.

43 Fred Krupp and Miriam Horn. *Earth: The Sequel.* New York: W.W. Norton & Company, Inc., 2008. p. 25.

44 Ibid., p. 26.

45 Bill Tomson. "U.S. plan to tolerate unapproved GMOs in crops draws concern." Dow Jones Newswires, Feb 2, 2009.

46 Brooke Jarvis. "The page that counts." *Yes! Magazine.* Spring 2008. www.yesmagazine.org/issues/climate-solutions/the-page-that-counts

47 George Leopold. "Demand soars for energy efficient electronics." *EE Times,* Feb 17, 2009, www.eetimes.com/news/latest/showArticle.jhtml?articleID=214303485

48 Roberts.

49 See Jean-Marie Macabrey, "Soaring electricity use by new electronic devices imperils climate change efforts." The *New York Times,* May 14, 2009. www.nytimes.com/cwire/2009/05/14/14climatewire-soaring-electricity-use-by-new-electronic-de-12208.html

50 Shahid Naeem."Lessons from the reverse engineering of nature." *Miller-McCune* magazine, May/June 2009. pp. 57–71.

CHAPTER 4: LISTEN FOR THE JAGUAR

1 Patrick Luganda. "Trees getting bigger absorbing more carbon." Network of Climate Journalists of the Greater Horn of Africa website, Mar 11, 2009. www.necjogha.org/news/2009-03-11/trees-getting-bigger-absorbing-more-carbon

2 Stephen Leahy. "North American trees dying twice as fast." Inter Press Service News Agency, Jan 22, 2009. ipsnews.net/news.asp?idnews=45511

3 "One-fifth of fossil-fuel emissions absorbed by threatened forests." *ScienceDaily,* Feb 19, 2009. www.sciencedaily.com/releases/2009/02/090218135031.htm

4 Leahy. "North American trees dying twice as fast."

5 Quoted in "One-fifth of fossil-fuel emissions absorbed by threatened forests." Emphasis added.

6 Interview by the authors, Mar 2009.

7 See "The Natural Step." Collins Pine website. www.collinswood.com/HistoryPhilosophy/TheNaturalStep.html

8 Klemens Laschefski and Nicole Freris. "Saving the wood from the trees." *The Ecologist,* July/Aug 2001. pp. 40–43.

9 Florianne Koechlin. "The dignity of an oak leaf." In *Cell Whispers: Journeys through New Realms of Science.* Basel: Lenos Verlag, 2005. www.gmo-free-regions.org/fileadmin/files/gmo-free-regions/Food_and_Democracy/Dignity_of_Oakleaf.pdf

10 See Paul Jay, "The beetle and the damage done," CBC News, Apr 23, 2008; www.cbc.ca/news/background/science/beetle.html

11 Koechlin.

12 Quoted in Carl Zimmer, "The web below." *Discover,* Nov 1, 1997. http://discovermagazine.com/1997/nov/thewebbelow1271

13 Arthur Partridge. "Mountain pine beetles—another look." *Ecoforestry.* Vol. 17, Number 1, 2002. pp. 4–8.

14 Ben Parfitt. *Over-cutting and Waste in B.C.'s Interior: A Call to Rethink B.C.'s Pine Beetle Logging Strategy.* Ecojustice Canada, June 2007. p. 15. www.ecojustice.ca/publications/reports/over-cutting-and-waste-in-b-c-s-interior/attachment. Emphasis added.

15 Established in 2005 by the International Network for Bamboo and Rattan (INBAR) and the International Fund for Agricultural Development (IFAD). "Introduction." *Global NTFP Partnership Strategy Document.* http://ntfp.inbar.int/wiki/index.php/About

16 "Activities—participatory forest management." Non-Timber Forest Products and Participatory Forest Management website. http://forests.hud.ac.uk/part_management.htm

17 Susan Zwinger. "The wisdom of an eco-forester." Interview with Merv Wilkinson, Apr 12, 1994.

18 Interview by Holly Dressel, Summer 2001; Mar 2009.

19 Andy White and Alejandra Martin. *Who Owns the World's Forests? Forest Tenure and Public Forests in Transition.* Washington, DC: Center for International Environmental Law, 2002. p. 2. www.forest-trends.org/documents/files/doc_159.pdf

20 Garrett Hardin. "The tragedy of the commons." *Science,* 162, 1968. pp. 1243–48; B. McCay and J.A. Acheson, eds. *The Question of the Commons: the Culture and Ecology of Communal Resources.* Tucson: University of Arizona Press, 1988. p. xiii.

21 Quoted in Scott Atran, "Itza Maya tropical agro-forestry." *Current Anthropology,* Vol. 34, 1993. pp. 633–700.

22 "Introduction to adaptive management." Coastal First Nations. http://coastalfirstnations.ca/files/PDF/EBM/2/AMbackgrounder.pdf. Emphasis added.

23 Ibid.

24 See "Ecosystem-based management learning forum: Introduction to land use objectives." Coastal First Nations. http://coastalfirstnations.ca/files/Documents/LegalObjectivesBackgrounder.pdf

25 Gordon Hamilton. "Great Bear forest deal shifts power after years of grinding negotiation." *Vancouver Sun,* Apr 1, 2009. www.sierraclub.bc.ca/quick-links/media-centre/media-clips/great-bear-forest-deal-shifts-power-after-years-of-grinding-negotiation/

26 "Traceable timber amendments strengthen Europe's illegal logging proposals." w w f's g f t n website, Feb 17, 2009. http://gftn.panda.org/?156601/ e u-Parliament-committee-passes-test-on-illegal-logging-law

27 "Protected forests crucial to supplying the world's biggest cities with cheaper clean water." World Bank-w w f Alliance for Forest Conservation. Sept 1, 2003; www.panda.org/about_our_earth/about_forests/ forest_news_resources/?8547/Protected-forests-crucial-to-supplying-the-worlds-biggest-cities-with-cheaper-clean-water

28 James Astill. "Cameroon's 'protected' forest is a meal ticket for elephant poachers." The *Guardian,* Aug 7, 2001. www.guardian.co.uk/world/2001/ aug/07/unitednations

29 Ibid.

30 "Traceable timber amendments strengthen Europe's illegal logging proposals."

31 "Environmentally and socially responsible trade financing mechanisms key to conserving world's forests." w w f's g f t n website, Mar 16, 2009. http://gftn.panda.org/?159081/Environmentally-and-Socially-Responsible-Trade-Financing-Mechanisms-Key-to-Conserving-Worlds-Forests

32 "Congo Basin passes 1 million ha milestone in swing to sustainable forestry." w w f's g f t n website, July 31, 2008. www.panda.org/what_we_do/successes/ ?14570 8/Congo-Basin-passes-1-million-ha-milestone-in-swing-to-sustainable-forestry

33 "U.S. cracks down on illegal logging." w w f's g f t n website, Oct 22, 2008. www.gftn.panda.org/newsroom/news_north_america.cfm?148541/ u s-cracks-down-on-illegal-logging/

34 Simon Lewis quoted in "One-fifth of fossil-fuel emissions absorbed by threatened forests."

CHAPTER 5: A RIVER RUNS THROUGH IT

1 Interview by Holly Dressel, Mar 2009.

2 "Immovable Maude: The life and times of Maude Barlow." *Life and Times.* c b c Television, Oct 9, 2001. www.cbc.ca/lifeandtimes/barlow.html

3 See bibliography in Maude Barlow, *Blue Covenant: The Global Water Crisis and the Coming Battle for the Right to Water.* Toronto: McClelland & Stewart, 2007. pp. 209–11.

4   "Water crisis hits West, South Delhi." *The Hindu*, Jan 4, 2005.
     www.thehindu.com/2005/01/04/stories/2005010415010300.htm; Urvashi Gulia.
     "Water riots: 35 areas identified." The *Times of India*. March 26, 2004.
     http://timesofindia.indiatimes.com/city/delhi/Water-riots-3areas-identified/
     articleshow/584664.cms

5   Allerd Stikker. "Water today and tomorrow: Prospects for overcoming scarcity."
     Futures, Vol. 30, Number 1, Jan 1998.

6   Maude Barlow and Tony Clarke. *Blue Gold: The Battle against Corporate Theft
     of the World's Water.* London: Earthscan, 2003. p. 16.

7   Someshwar Singh. "The covert ground-water crisis." Third World Network,
     Mar 3, 2000. www.twnside.org.sg/title/covert.htm. Emphasis added.

8   Madhu Kishwar."Villages in Rajasthan overcome Sarkari dependence,"
     Manushi, Number 123, Apr 22, 2009. Available at www.indiatogether.org/
     manushi/issue123/rajendra.htm

9   Daniel Pepper. "India's water shortage." *Fortune*, Jan 29, 2009.
     http://money.cnn.com/2008/01/24/news/international/India_water_shortage.
     fortune/index.htm

10   "Tarun Bharat Sangh is transforming rural Rajasthan by awakening old
     memories!" *Good News India* website. www.goodnewsindia.com/Pages/
     content/inspirational/tbs.html

11   See the Government of India Water Resources website at http://wrmin.nic.in/
     newdetails.asp?langid=1&newscode=14

12   Marq de Villiers. *Water.* Toronto: Stoddart, 1999. p. 171.

13   Priit J. Vesilind. "The Middle East's Water: Critical resource."
     National Geographic, Vol. 183, Number 5, May 1993.

14   Ibid.

15   Ibid.

16   De Villiers, p. 171.

17   A subsection of the Canadian mining act has even permitted lakes to be
     reclassified as tailing dumps! See Terry Milewski, "Lakes across Canada
     face being turned into mine dump sites." C B C News, June 16, 2008.
     www.cbc.ca/canada/story/2008/06/16/condemned-lakes.html

18   Gwynne Dyer. *Climate Wars.* Toronto: Random House Canada, 2008. p. 9.

19   See "Rainwater harvesting in dry lands."
     www.rainwaterharvesting.org/international/dryland.htm

20 "Green infrastructure." North Carolina Division of Forest Resources website.
    http://dfr.nc.gov/Urban/urban_green_infrastructure.htm

21 "Poisoned waters." *Frontline*. PBS, Apr 21, 2009.
    www.pbs.org/wgbh/pages/frontline/poisonedwaters/

22 Emily Pilloton. "Chicago green roof program." Inhabitat website, Aug 1, 2006.
    www.inhabitat.com/2006/08/01/chicago-green-roof-program/

23 Liat Podolsky. *Green Cities, Great Lakes: The Green Infrastructure
    Report*. Ecojustice, Sept 2008. www.ecojustice.ca/publications/reports/
    the-green-infrastructure-report/attachment

24 John McNeil. *Oakville's Urban Forest: Our Solution to Our Pollution*.
    Oakville, ON: Corporation of the Town of Oakville, 2006.
    www.itreetools.org/resource_learning_center/elements/CityTrees_Mar_
    Apr07_Oakville article.pdf

25 Article IV. *Statement of the Fund Mission*. Ottawa, Dec 7, 1995. pp. 16–17.
    Obtained by the Halifax Initiative under the *Access to Information Act*.
    Sept 30, 1999. http://halifaxinitiative.info/content/imfs-structural-adjust-
    ment-programme-canada-1994-1995-december-1995. Emphasis added.

26 For a detailed discussion about how IMF structural adjustment has affected
    Canadian social and especially health programs, see Holly Dressel,
    *Who Killed the Queen: The Story of a Community Hospital and How to Fix
    Public Health Care*. Montreal: McGill-Queen's University Press, 2008.
    pp. 287–96.

27 Ish Theilheimer. "Downsized water expert finds no profit in protecting
    Walkerton residents." *Straight Goods*, May 31, 2000.

28 See Maude Barlow. "Blue gold: The global water crisis and the commodification
    of the world's water supply." Special report issued by the International Forum
    on Globalization, Spring 2001.

29 Roberta Avery and Kate Harris. "Stories of Walkerton."
    Toronto Star, May 20, 2001.

30 Robert Goodland, Laura Orlando, and Jeff Anhang. *Toward Sustainable
    Sanitation*. Fargo, ND: The International Association of Impact Assessment,
    May 2001.

31 Carola Vyhnak. "Soiled land," Sludge Series. *Toronto Star*, July 12–15, 2008;
    "Tainted soil in Milwaukee to be removed." *Journal Sentinel*, Aug 6, 2007.
    www.biosolids.org/news_weekly.asp?id=2028; Maureen Reilly. "Spreading
    sludge on farms risks health, environment." *Hamilton Spectator*, June 19, 2008.

www.thespec.com/printArticle/389078; Lishanti Caldera. "Recycled paper sludge leaching into environment."
Epoch Times, Aug 9, 2007. en.epochtimes.com/tools/printer.asp?id=58571

32 Besides the obvious danger of live *Salmonella, E. coli,* and every imaginable communicable disease, bacteria, and virus, there are numerous examples of bio-accumulation in which cows, wild geese, and earthworms sustained toxic levels of copper or heavy metals like thallium by grazing in sludged fields. Tests on vegetables have revealed mercury levels as much as fifty times the acceptable level, and lawsuits alleging everything from prolonged nausea and diarrhea to Crohn's Disease, malfunctioning kidneys, and elevated levels of lead and barium are being found in residents near sludge-spread fields or lagoons. Many medical groups, such as the Canadian Infectious Disease Society, the National Academy of Sciences, the Centers for Disease Control and Prevention, Cornell Waste Management Institute, and Johns Hopkins University, to name just a few, have demanded moratoriums on spreading sludge on agricultural land because of the associated health risks.

33 Quebec Superior Court decision, case no. 760-17-001371-064.
Judge Steve J. Reimnitz presiding, Oct 1, 2009. It is now being challenged by the sludge industry.

34 Quoted in Berklee Lowrey-Evans, "Hope and renewal in the Iraqi marshlands." International Rivers website, Dec 15, 2008. www.internationalrivers.org/en/node/3662

35 "China warns of environmental 'catastrophe' from Three Gorges Dam." *Xinhua News,* Sept 26, 2007. http://news.xinhuanet.com/english/2007-09/26/content_6796234.htm

36 Jim Yardley. "Chinese dam projects criticized for their human cost." The *New York Times,* Nov 19, 2007; Antoaneta Bezlova. "China reins in dam builders." Inter Press Service News Agency, June 18, 2009. http://ipsnews.net/print.asp?idnews=47259

37 Elizabeth Brink. "Feeding a hungry river." International Rivers website, Dec 15, 2008. www.internationalrivers.org/en/node/3638/

38 "Ilisu dam fighters honored at award ceremony in Istanbul." International Rivers press release, Mar 21, 2009. http://internationalrivers.org/en/node/4093

39 "Frequently asked questions." International Rivers website. www.internationalrivers.org/en/node/480

40 Arundhati Roy. *Power Politics, Second Edition.* Cambridge, M A: South End Press, 2001. p. 44.

41 Emanuele Lobina. "Water privateers, out!" *Focus on the Public Services,* Feb 2000.

42 Ibid.

43 Emanuele Lobina and David Hall. "Public sector alternatives to water supply and sewerage privatisation: Case studies." *International Journal of Water Resources Development,* Vol. 16, Number 1, 2000. pp. 35–55. www.psi-jc.jp/news_policy/policy/water_in_public_hands/9908--W-U-Pubalt.doc. In June 2009 S A N A A employees, under increasing pressure to provide city water because of a failed wastewater plant donated by the EU, joined demonstrations to protest the coup against democratically elected President Manuel Zelaya; see Berta Joubert-Ceci, "General strike resists Honduras coup." *Workers World,* July 1, 2009. www.workers.org/2009/world/honduras_0709/

44 See *The Way Forward: Public Sector Water and Sanitation.* P S I briefing to the World Water Forum, The Hague, Mar 17–22, 2000.

45 Olivier Hoedeman. "Europe seeks moratorium to protect public water from privatization push." AlterNet website, Nov 1, 2008. www.alternet.org/module/printversion/105465

46 "Financial collapse builds case for public services." *CUPE* website, Oct 16, 2008. http://cupe.ca/priwatchoct08/Financial-collapse-b

47 This covenant is what Maude Barlow and all the new watchdogs listed in this section are fighting for at the UN level. As they are making progress, it is not unrealistic to imagine that a country basing its policies on privatization could be caught in a difficult situation. See Maude Barlow, *Blue Covenant.* Toronto: McClelland and Stewart, 2007. p. 165.

48 Quoted in Barlow, *Blue Covenant.* p. 175.

49 Quoted in Hilmi Toros, "Troubled waters hard to bridge." Report on the Fifth World Water Forum, Istanbul. Inter Press Service News Agency, Mar 22, 2009. www.ipsnews.net/print.asp?idnews=46227

50 Barlow. *Blue Covenant.* p. 174.

51 "Enforcement and litigation." Riverkeeper website. www.riverkeeper.org/about-us/our-methods/enforcement-litigation/

52 Barlow and Clarke. p. 221.

53 Interview by Kristin Palitza. "Development must adapt to water resources we have." Inter Press Service News Agency, Feb 3, 2009. http://ipsnews.net/news.asp?idnews=45658

CHAPTER 6: THE MOTHER OF ALL

1   Alanna Mitchell. *Sea Sick: The Global Ocean in Crisis*. Toronto: McClelland & Stewart, 2009. p. 8.

2   Ibid, pp. 14 and 22.

3   Ibid, p. 17.

4   Ibid, p. 35.

5   Interview by David Suzuki, 2001.

6   Interview for The Suzuki Diaries documentary series. "Coastal Canada." *The Nature of Things*. CBC Television, Nov 22, 2009. www.cbc.ca/documentaries/natureofthings/2009/suzukidiaries/

7   Christopher Dyer and Richard Leard in E. Pinkerton and M. Weinstein, eds. *Fisheries That Work: Sustainability through Community-Based Management*. Vancouver: The David Suzuki Foundation, 1995. p. 101.

8   Quoted in Larry Pynn, "Special edition," a five-part series on marine conservation. *Vancouver Sun,* Apr 30–May 4, 2001.

9   Joachim Dyfvermark and Fredrik Laurin. "The illegal cod." Report for TV4, Sweden, 2006.

10 Callum Roberts and Julie Hawkins. *Fully Protected Marine Reserves: A Guide*. York, UK: University of York, 2000.

11 Larry Pynn, "Special Edition."

12 Ibid. Emphasis added.

13 Ibid.

14 Jane George, "Feds to protect Fury and Hecla Strait in Nunavut," Oct 29, 2009. www.nunatsiaguonline.ca/stories/article/724_feds_float_plan_to_protect_fury_and_hecla_stait/

15 George Monbiot. "These are not the mariners of old but pirates who make bureaucrats blanch." The *Guardian,* June 1, 2009. www.guardian.co.uk/commentisfree/2009/jun/01/george-monbiot-marine-fisheries-law

16 Filmed interview by filmmaker Ryan Young for the upcoming documentary *Seal Wars*, Nov 2008.

17 See the press release on the occasion of Paul Watson receiving the Amazon Peace Prize, July 12, 2007: www.seashepherd.org/news-and-media/news-070712-1.html. For anti–Sea Shepherd bias, see the many articles on the seizure of the *Farley Mowat* in April 2008 on cbc.ca, culminating with July 1, 2009, "Anti-sealing activists convicted," where the report notes that none of the defendants appeared and neglects to mention that they had all been denied visas to enter Canada to do so, as well as their claim to have been within legal distance of the killing activities. www.cbc.ca/Canada/nova-scotia/story/2009/07/01/ns-sealing-trial.html

CHAPTER 7: BAKING A SUSTAINABLE PUDDING

1 "The first Malian Ark product."Slow Food Foundation for Biodiversity website, Oct 18, 2007. www.slowfoodfoundation.com/eng/leggi.lasso?cod=3E6E345B18 daf2C81FUHYl1EF071&ln=en

2 Interview in David Suzuki and Holly Dressel, *From Naked Ape to Superspecies.* Vancouver: Greystone Books/David Suzuki Foundation, 2004. pp. 337–38.

3 "The journey of the Ark of Taste continues." *Slow Food Foundation for Biodiversity* website, May 25, 2009. www.slowfoodfoundation.com/eng/leggi.lasso?cod=3E6E345B0cd1d2869ATyu2FA6ADD&ln=en

4 Sandra Messick. "Cadbury bites into ethics." The *Independent,* Mar 6, 2009. Reproduced at www.slowfood.com/sloweb/eng/dettaglio.lasso?cod=3E6E345 B077911B855TJy26BA781; "Mars, Inc. announces sustainability commitments for cocoa." International Labor Rights Forum website, Apr 10, 2009. www.laborrights.org/stop-child-labor/cocoa-campaign/news/11854

5 "Defending biodiversity. " Slow Food and Terra Madre website. Newsletter. http://slowfood.com/slowfood_time/15/eng.html#B

6 Slow Food Foundation for Biodiversity. *Slow Food Presidia.* Bra, Italy: Slow Food, 2009. www.slowfoodfoundation.com/pdf/pubblicazioni/sf_presidi_eng.pdf

7 See the report of the French sanitary agencies (in French): www.invs.sante.fr/publications/2004/inf_origine_alimentaire/grilleLecture.pdf; see also "Food safety and foodborne illness," World Health Organization, March 2007. www.who.int/mediacentre/factsheets/fs237/en/

8   Excerpt taken from Woody Tasch, *Inquiries into the Nature of Slow Money:*
      Investing as if Food, Farms, and Fertility Mattered. White River Junction,
      VT: Chelsea Green Publishing, 2008. p. 6.

9   Ibid., p. 119.

10   Quoted in Roger Bybee, "Growing Power in an urban food desert."
      Yes! Magazine, Spring 2009. pp. 28–31. www.yesmagazine.org/issues/
      food-for-everyone/growing-power-in-an-urban-food-desert

11   Norman Borlaug. "We need biotech to feed the world." *Wall Street Journal*
      editorial, Dec 6, 2000.

12   Bill Freese. "Biotech snake oil: A quack cure for hunger."
      Multinational Monitor, Vol. 29, Number 2, Sept/Oct, 2008.
      www.multinationalmonitor.org/mm2008/092008/freese.html

13   See "Farmers get better yields from new drought-tolerant cassava."
      International Institute for Tropical Agriculture (Nigeria) press release,
      Nov 3, 2008. www.iita.org/cms/details/news_details.aspx?articleid=
      1897&zoneid=81; "The sisters of nutrition." IRRI press release, Apr 3, 2000.
      www.irri.org/Hunger/Nutrition; "Conventional soybeans offer high yields
      at lower cost." University of Missouri press release, Sept 8, 2008.
      http://agebb.missouri.edu/news/ext/showall.asp?story_num=4547&iln=49

14   See Martin Webber, "GM crops 'may give lower yields.' " BBC World Service,
      Feb 6, 2009. http://news.bbc.co.uk/2/hi/business/7866687.stm

15   Interview by the authors, Fall 2009.

16   Javier Blas. "The end of abundance: food panic brings calls for a second
      'green revolution.'" *Financial Times,* June 1, 2008.
      www.ft.com/cms/s/0/1b6fd476-2ff3-11dd-86cc-000077b07658.html

17   "Choose and take what you want," *Cumhuriyet,* June 13, 2009.
      http://haber.turk.net/ENG/2288097/Turkey-Press-Scan--2-

18   "Africa becomes wary of farm deals: land activist." Food Crisis and the Global
      Land Grab website, June 9, 2009. farmlandgrab.org/date/2009/06/page/5;
      Olivier Bourque. "Quebec: Des investisseurs chinois lorgnent des terres
      agricoles." *Argent,* Sept 25, 2009. argent.canoe.ca/lca/affaires/quebec/arch
      ives/2009/09/20090925-165750.html

19   "Food security in a volatile world." IAASTD *Issues in Brief.* Washington, DC:
      Island Press, 2008. www.agassessment.org/docs/10505_FoodSecurity.pdf.
      Emphasis added.

20   Ibid.

21 "Multifunctional agriculture for social, environmental and economic sustainability." I A A S T D *Issues in Brief.* Washington, D C: Island Press, 2008.
www.agassessment.org/docs/10505_multi.pdf

22 "Executive summary of the synthesis report." *Agriculture at a Crossroads.*
Washington, D C: I A A S T D, 2008. www.agassessment.org/reports/I A A S T D/E N/
Agriculture%20at%20a%20Crossroads_Executive%20Summary%20of%20
the%20Synthesis%20Report%20%28English%29.pdf

23 "Jonas." Posting in response to "Agriculture produces more than just crops—
and it's time for policy to reflect that." *Grist* website, 9:49 A M,
May 22, 2008. www.grist.org/article/farm-and-function

24 See, for example, the world ecological footprint calculated at www.nation-
master.com/graph/env_eco_foo-environment-ecological-footprint

25 Quoted in "Inter-governmental report aims to set new agenda for
global food production." I A A S T D press release, Mar 31, 2008.
www.agassessment.org/docs/I A A S T D_backgroundpaper_280308.doc.

26 Interview by Holly Dressel, Winter 2010.
See also Bruce Bridgeman. "Stopping population growth essential." *Santa Cruz
Sentinel,* Mar 29, 2009. www.santacruzsentinel.com/localstories/ci_12022164

27 Quoted in Dr. Mae-Wan Ho. "Organic Cuba without fossil fuels."
Institute of Science in Society website, Jan 21, 2008.
www.i-sis.org.uk/OrganicCubawithoutFossilFuels.php

28 Ibid.

29 See "Status of women," Government of Kerala website. www.kerala.gov.
in/education/status.htm. Canada's birth rate is given in "Women over 30
help birthrate rise in 2007." C B C News, Sept 22, 2009. www.cbc.ca/canada/
story/2009/09/22/canada-births-2007-statistics-boom.html

30 Quoted in Charles C. Mann, "Our good earth." *National Geographic,* Sept 2008.
pp. 97–100. ngm.nationalgeographic.com/print/2008/09/soil/mann-text

31 Ibid.

32 Lawrence Alderson. *Rare Breeds.* New York: Little, Brown & Co., 1994. p. 79.

33 Joel Salatin. *Everything I Want to Do Is Illegal.* Swoope, V A: Polyface, Inc.,
2007. p. 144.

34 See, for example, "A food system that kills." G R A I N website, Apr 2009.
www.grain.org/articles/?id=48

35 Interview by Holly Dressel, Fall 2009.

36 Interview by Holly Dressel, Spring 2001.

37 See "Avian influenza goes global, but don't blame the birds," *The Lancet Infectious Diseases,* Vol. 6, Apr 2006. p. 185. GRAIN is a small international nonprofit organization that works to support small farmers and social movements in their struggles for community-controlled and biodiversity-based food systems.

38 See R.J. Garten, et al. "Antigenic and genetic characteristics of swine-origin 2009 A (H1N1) influenza viruses circulating in humans." *Science,* May 22, 2009; see also G.M. Nava, et al. "Origins of the new Influenza A (H1N1) virus: Time to take action." *Eurosurveillance,* Vol. 14, Number 22, June 4, 2009. www.eurosurveillance.org/ViewArticle.aspx?ArticleId=19228

39 A recent article in *Environmental Health Perspectives* makes it clear that under normal conditions, such viruses would "burn out, whereas in CAFOS [concentrated animal feeding operations], which often have continual introductions of [unexposed] animals—there's a much greater potential for the viruses to spread and become endemic." See Charles W. Schmidt, "Swine CAFOS and novel H1N1 flu." *Environmental Health Perspectives,* Vol. 117, Number 9, Sept 2009. p. A396.

40 "A food system that kills."

41 Michael Pollan. "Farmer in chief." The *New York Times,* Oct 12, 2008. www.nytimes.com/2008/10/12/magazine/12policy-t.html

42 Marcia Ishii-Eiteman of the Pesticide Action Network, for example, said, "Future conditions will not be like in the past. All bets are off. We need to focus on creating adaptive, resilient farming systems." Quoted in Stephen Leahy, "Factoring farming into climate change," Inter Press Service News Agency, Apr 2, 2009. www.straightgoods.ca/2009/ViewFeature.cfm?Ref=228

43 Pollan.

44 "Episode 1." *ReGeneration,* interview by Claude William Genest. Vermont Public Television (PBS). Transcript at www.vpt.org/programs/regen_collins.html

45 Salatin. pp. 133–35. Emphasis added.

46 Quoted in "Episode 1." *ReGeneration.*

47 Eric Holt-Gimanez, quoted in Miguel A. Altieri and Parviz Koohafkan, *Enduring Farms: Climate Change, Smallholders and Traditional Farming Communities.* Panang, Malaysia: Third World Network, 2008. www.fao.org/fileadmin/templates/giahs/PDF/Enduring_Farms.pdf . Emphasis added.

48 Sandra Steingraber. "The ecology of pizza (or why organic food is a bargain)."
    From the Organic Trade Association, June/July 2006.

49 Charles C. Mann. *1491: New Revelations of the Americas before Columbus.*
    New York: Knopf, 2005. p. 298.

50 "ICIPE announces safe new methods for controlling stemborers, termites
    and striga." ICIPE press release, Nairobi, Kenya, June 15, 2000.

CHAPTER 8: GREEN JOBS IN THE CITY

1   Interview by Holly Dressel, Fall 2009.

2   Interview by Holly Dressel, Spring 2002.

3   "Vision for the Center for Earth Leadership."
    Center for Earth Leadership website. www.earthleaders.org/about

4   Ibid. Emphasis added.

5   See, for example, Democracy Watch (www.dwatch.org) and Corporate Watch
    (www.corpwatch.org)

6   "Cleveland goes to Mondragón." *Owners at Work,* Winter 2008–09. pp. 10–12.

7   Ibid.

8   From description of Mondragón, "Mondragon Corporacion Cooperativa, Spain,"
    on International Institute for Sustainable Development website.
    www.iisd.org/50comm/commdb/list/c13.htm

9   See under "What is the secret of the Mondragon Experience's success?"
    in Mondragón's list of frequently asked questions on its website.
    www.mondragon-corporation.com/language/en-US/ENG/Frequently-
    asked-questions/Corporation.aspx

10 "All in this together." The *Economist,* Mar 26, 2009.
    www.economist.com/businessfinance/displayStory.cfm?story_id=
    13381546&source=login_payBarrier

11 See under "How many of your employees are cooperative members and
    how many are not? What areas do the non-members usually work in?"
    in Mondragón's list of frequently asked questions on its website.
    www.mondragon-corporation.com/language/en-US/ENG/Frequently-
    asked-questions/Corporation.aspx

12 Tim Huet. "Can coops go global?" *Dollars and Sense,*
    Nov/Dec 1997. www.dollarsandsense.org/archives/1997/1197huet.html

13 Michael Fox. "Mercosur's cooperatives in an age of integration."
    Third World Resurgence, Number 224, Apr 2009. p. 36.
    www.twnside.org.sg/title2/resurgence/2009/224/actions1.htm

14 Huet.

15 Ibid.

16 Ibid.

17 Wendell Berry. "Conserving communities." In *Another Turn of the Crank*.
    Berkeley, CA: Counterpoint Press, 1995. p. 20.

18 Holly Sklar. "No foreclosures here." *Yes! Magazine,* Winter 2009. pp. 51–54.
    www.yesmagazine.org/issues/sustainable-happiness/no-foreclosures-here

19 See Wuppertal Institute website at www.wupperinst.org/en

20 Interview by Holly Dressel, Winter 2001.

21 "Bolivarian Revolution in Peru." Entry in Caracas Gringo blog,
    June 16, 2009. http://caracasgringo.wordpress.com/2009/06/16/
    bolivarian-revolution-in-peru/

22 "'New South America': old lies in new bottles."
    International Communist Current (ICConline), May 4, 2009.
    http://en.internationalism.org/icconline/2009/05/new-sa

23 Anastasia Moloney. "The challenge of South America's populist left."
    World Politics Review, Jan 12, 2009.
    www.worldpoliticsreview.com/article.aspx?id=3146

24 Interview by Holly Dressel, Fall 2009.
    See also "Benefits of export taxes." Third World Network preliminary paper.
    www.twnside.org.sg/title2/par/Export_Taxes.doc.

25 Interview; See also T.J. Marta and Joseph Brusuelas. *Forex Analysis and
    Trading.* Bloomberg Press, 2009.

26 "Morales to shake free of IMF yoke."
    Inter Press Service News Agency, Apr 20, 2006. Available at
    www.globalexchange.org/countries/americas/bolivia/3933.html

27 Johann Hari. "Is the U.S. about to treat the rest of the world better? Maybe…"
    The *Independent,* Jan 23, 2009. www.independent.co.uk/opinion/
    commentators/johann-hari/johan-hari-is-the-us-about-to-treat-the-
    rest-of-the-world-better-maybe-1513367.html

28 Bernard Lietaer. "Beyond greed and scarcity." Dialogue with
    Sarah van Gelder. *Yes! Magazine,* Spring 1997. www.yesmagazine.org/issues/
    money-print-your-own/beyond-greed-and-scarcity

29 David Korten. "Money versus wealth." *Yes! Magazine,* Spring 1997. www.yesmagazine.org/issues/money-print-your-own/money-versus-wealth/?searchterm=money%20versus%20wealth

30 Lietaer. Emphasis added.

31 See the Triodos website at www.triodos.com/com/about_triodos/

32 Quoted in Madeline Ostrander, "Life reclaimed." *Yes! Magazine,* Winter 2009, p. 28. www.yesmagazine.org/issues/sustainable-happiness/life-reclaimed

33 Jeanna Bryner. "Key to happiness: Give away money." *Live Science,* Mar 20, 2008. www.livescience.com/health/080320-happiness-money.html

34 Sharon Begley. "Why money doesn't buy happiness." *Newsweek,* Oct 15, 2007. www.newsweek.com/id/43884

INDEX

THE DAVID SUZUKI FOUNDATION

The David Suzuki Foundation works through science and education to protect the diversity of nature and our quality of life, now and for the future.

With a goal of achieving sustainability within a generation, the Foundation collaborates with scientists, business and industry, academia, government and non-governmental organizations. We seek the best research to provide innovative solutions that will help build a clean, competitive economy that does not threaten the natural services that support all life.

The Foundation is a federally registered independent charity that is supported with the help of over 50,000 individual donors across Canada and around the world.

We invite you to become a member. For more information on how you can support our work, please contact us:

The David Suzuki Foundation
219–2211 West 4th Avenue
Vancouver BC Canada v6K 4s2
contact@davidsuzuki.org
Tel: 604-732-4228
Fax: 604-732-0752

Checks can be made payable to The David Suzuki Foundation. All donations are tax-deductible.

Canadian charitable registration: (BN) 12775 6716 RR0001
U.S. charitable registration: #94-3204049